AF294786

Discrete Structural Optimization

International Union of Theoretical
and Applied Mechanics

W. Gutkowski · J. Bauer (Eds.)

Discrete Structural Optimization

IUTAM Symposium Zakopane, Poland
August 31 – September 3, 1993

Springer-Verlag

Berlin Heidelberg New York
London Paris Tokyo
Hong Kong Barcelona Budapest

Prof. Witold Gutkowski
Dr. Jacek Bauer

Institute of Fundamental Technological Research
ul. Swietokrzyska 21
PL-00-049 Warsaw
Poland

ISBN-13:978-3-642-85097-4   e-ISBN-13:978-3-642-85095-0
DOI: 10.1007/978-3-642-85095-0

Typesetting: Camera ready by authors
SPIN: 10083220   61/3020 – 5 4 3 2 1 0 – Printed on acid-free paper.

Preface

The IUTAM Symposium on "Discrete Structural Optimization" was devoted to discuss optimization problems for which design variables may not be sought among continuous sets. Optimum sizing from lists of available profiles, segmentation; as well as allocation and number of supports, sensors or actuators, are good examples of such problems for which design variables may be chosen only from finite sets. The above problems are having not only important practical applications. They are also inspiring scientific research in the field of discrete applied mathematics. Among them are controlled enumeration methods, subgradient approach, genetic programming, multicriteria optimization, neural nets etc.

Along with its tradition of promoting and supporting new important fields of research in mechanics and its application, General Assembly of IUTAM decided in 1990 to support the Sumposium. It is worthy to note that this is the second IUTAM Symposium on structural optimization, organized in Poland. The first one was held in Warsaw twenty years ago and was organized by Professors Sawczuk and Mróz. It was devoted mostly to problems with continuous design variables.

The Symposium which gathered 40 participants from 12 countries was sponsored by several institutions listed below. However support by IUTAM should be specially appreciated. It helped several scientists to contribute to the Symposium which other way wouldn't attend the meeting.

Finally let me express my appreciations to members of Scientific Committee, members of Local Organizing Committee for their contribution for the successful scientific meeting. Thanks also to Springer - Verlag for very active cooperation in publishing this book.

December 1993

Witold Gutkowski
Symposium Chairman

Scientific Committee

B. A. Boley, U.S.A.
G. Cheng, China
H. Eschenauer, F.R.G.
C. Fleury, Belgium
D. Grierson, Canada
W. Gutkowski, Poland, (Chairman)
R. I. Haftka, U.S.A.
J. Koski, Finland
A. B. Templeman, U.K.
P. Trompette, France
G. N. Vanderplaats, U.S.A.
Y. Yamamoto, Japan

Local Organizing Committee

W. Gutkowski, Chairman
J. Bauer, Secretary
Z. Iwanow
D. Wincewicz
J. Zawidzka

Sponsoring Institutions

International Union of Theoretical and Applied Mechanics
Inst. of Fundamental Technological Research, Warsaw, Poland
Mostostal - Export S.A., Poland
Committee for Mechanics, Polish Academy of Sciences.

List of Participants

Dr. J. Bauer	Inst. of Fundamental Technological Research, ul. Swiętokrzyska 21, Warszawa, POLAND
Dr. Ahmed A. Belal	Faculty of Engng., University of Alexandria, 37 Sorya st. Roushdy Alexandria, EGYPT
Prof. Algirdas Čižas	Vilnius Technical University, Sauletekio al.11, 2054 Vilnius, LITHUANIA
Dr. A. S. Dekhtjar	Academy of Art Architectural Division Smirnova - Latochkina str.,20, 252053 Kiev, UKRAINE
Prof. K. Dems	Politechnika Łódzka - I26, ul. Żwirki 36, 90-924 Łódź, POLAND
Prof. H. A. Eschenauer	University of Siegen, Paul-Bonatz - Str.9-11, W-5900 Siegen, GERMANY
Prof. M. B. Fuchs	Dept. of Solid Mechanics Material and Structures, Faculty of Engineering, Tel-Aviv University, 69978 Ramat-Aviv, ISRAEL
Dr. J. Geilen	Univ. of Siegen, Inst. of Mech. and Control Engng. IMR, Paul-Bonatz-Str. 9-11, D-5900 Siegen, GERMANY
Prof. W. Gutkowski	Inst. of Fundamental Technological Research, ul. Swiętokrzyska 21, Warszawa, POLAND
Dr. P. Hajela	Mech. Engng., Aeronaut. Engng. and Mech., Rensselaer Polytechnic Institute, Troy, NY 12180, USA
Prof. J. Holnicki-Szulc	Universidade da Beira Interior, Dept. Electromecanica Rua Marques d'Avila e Boloma, 6200 Covilha, PORTUGAL

VIII

Dr. A. Janczura — Inst. of Building Engng., Technical University of Wrocław, Pl. Grunwaldzki 11, 50-377 Wrocław, POLAND

Dr. S. Jendo — Inst. of Fundamental Technological Research, ul. Swiętokrzyska 21, Warszawa, POLAND

Dr. Mitsuo Kishi — Dptm. of Naval Architecture, University of Osaka Prefecture, 1-1, Gakuen-echo, Skai, Osaka 591, JAPAN

Prof. J. Korbicz — Dept. of Robotics and Software, Eng., Higher College of Eng., ul. Podgórna 50, 65-246 Zielona Góra, POLAND

Prof. J. F. Mączyński — Inst. of Fundamental Technological Research, ul. Swiętokrzyska 21, Warszawa, POLAND

Dr. J. Mottl — VABrno, Tererova 6, 61200 Brno, CZECH REPUBLIC

Prof. Z. Mróz — Inst. of Fundamental Technological Research, ul. Swiętokrzyska 21, Warszawa, POLAND

Dr. J. Niczyj — Instytut Inżynierii Lądowej, Politechnika Szczecińska, Al. Piastów 50, 70-311 Szczecin, POLAND

Prof. A. Osyczka — Politechnika Krakowska, Al. Jana Pawła II 37, 31-864 Kraków, POLAND

Dr. W. Paczkowski — Instytut Inżynierii Lądowej, Politechnika Szczecińska, Al. Piastów 50, 70-311 Szczecin, POLAND

Prof. E. P. Petrov — Shilovskij v'ezd 5/12a, 310013, Kharkov, UKRAINE

Dr. Mariusz Pyrz — Dept. Mecanique, Ecole Universitaire, D'ingenieurs de Lille, LML, USTL, (F-)59655 Villeneuve d'ASCQ, FRANCE

Dr. A. O. Rasskazov — Inst. of Automobiles and Highways, Suvorova str.,1, 252010 Kiev, UKRAINE

Prof. G. Rozvany — Fachbereich 10, Universitat Essen, Postfach 10 37 64, W-4300 Essen, GERMANY

Dr. E. Schäfer — Hammerstr. 7, 65594 Runkel (Ennerich), GERMANY

Mr. Serge Shevchenko — 6. Miroshnichenko str. app.1, Merefa, Kharkov region, UKRAINE, 312060

Prof. J. Sieczkowski — Inst. of Building Engng., Technical Univ. of Wrocław, Wyb. Wyspiańskiego 27, 50-370 Wrocław, POLAND

Dr. St. Stupak — Vilnius Technical University, Sauletekio al.11, 2054 Vilnius, LITHUANIA

Dr. V. Toropov — Dept. of Civil Engineering, Univ. of Bradford, Bradford West Yorkshire, BD 7 1 DP, UK

Prof. Ph. Trompette — Inst. de Mecanique de Grenoble, Domain Univ. B.P. n68, 38402 Saint-Martin-D'Heres, Cedex, FRANCE

Mrs. D. Wincewicz — Inst. of Fundamental Technological Research, ul. Swiętokrzyska 21, Warszawa, POLAND

Prof. Jiang Xiao yu — Automotive Engng. Depart., Tsinghua Univ., Beijing 100084, CHINA

Dr. N. Yoshikawa — Inst. of Industrial Science, Univ. of Tokyo, 7-22-11, Roppongi, Minato-ku, Tokyo 106, JAPAN

Mrs. J. Zawidzka — Inst. of Fundamental Technological Research, ul. Swiętokrzyska 21, Warszawa, POLAND

Dr. Ming Zhou — Essen University, FB 10, Postfach 10 37 64, D-4300 Essen 1, GERMANY

Scientific Program

Tuesday, September 31, morning

Opening of the Symposium

W. Gutkowski, Chairman

Session 1 : Genetic and Logical Structural Optimization
Chairman : H.A. Eschenauer

J.L. Marcelin, Ph. Trompette, C. Schmeding : *Genetic Optimization of Viscoelastically Damped Plates with Partial Coverage*

J. Mottl : *The Space Construction Optimization Composed from Wall Elements Using the "Voting Method"*

M. Kishi, T. Kodera, Y. Iwao, R. Hosoda : *Neuro-Optimizer, Its Application to Discrete Structural Optimization*

P. Hajela, E. Lee : *Genetic Algorithms in Topological Design of Grillage Structures*

J. Mączyński : *Optimization with a Linear Objective Function and Logical Constraints*

Tuesday, September 31, afternoon

Session 2 : Shape, arrangement, number of elements optimization
Chairman: Ph. Trompette

Z. Mróz, K. Dems : *On Optimal Structural Segmentation Problem*

E. Schäfer, H. A. Eschenauer, J. Geilen : *Application of Discrete Optimization Techniques to Optimal Composite Structures*

A. Osyczka, J. Montusiewicz : *A Method for Multicriterion Optimization of Discrete Nonlinear Models*

A.O. Rasskazov, A. C. Dekhtjar : *Optimal Arrangement of Ribs in Rib-reinforced Shells and Plates*

J. Niczyj, W. Paczkowski : *Application of the Expert System for Discrete Optimization of Space Truss*

Wednesday, September 1, morning

Session 3 : **Engineering Applications**
Chairman : Z. Mróz

V.V. Toropov, V. L. Markin, H. Carlsen : *Discrete Structural Optimization Based on Multipoint Explicit Approximations*

E.P. Petrov : *Optimization of Mistuning Parameters for Forced Vibration Stress Levels of Turbomachine Blades Assemblies*

Jiang Xiao yu, Chen Li xin : *Optimization of Gear Case Figure to Minimize the Radiated Noise*

Wednesday, September 1, afternoon

Session 4 : **Discretized Methods in Structural Optimization**
Chairman : A. Čižas

G.I.N. Rozvany, M. Zhou : *New Discretized Optimality Criteria Methods - State of the Art.*

G.I.N. Rozvany, M. Zhou, T. Birker, T. Lewiński : *Discretized Methods for Topology Optimization*

N. Yoshikawa, S. Nakagiri : *Homology Design of Flexible Structure by the Finite Element Method*

S. Shevchenko : *Optimization of Structure and Development of Production System*

Thursday, September 2, morning

Session 5 : **Sensors, Actuators and Supports Allocations**
Chairman : G.I.N. Rozvany

W. Gutkowski, J. Bauer, Z. Iwanow : *Support Number and Allocation for Optimum Structure*

D. Uciński, J. Korbicz : *Sensor Allocation for Parameter Identification of Two - Dimensional Distributed Systems*

J. Holnicki-Szulc, F. Lopez-Almansa, A. Mackiewicz : *Optimal Location of Piezoelectric Actuators*

Friday, September 3, morning

Session 6 : **Sizing**
Chairman : W. Gutkowski

A. Čižas, S. Stupak : *Optimal Discrete Design of Elastoplastic Structures*

S. Jendo, W. M. Paczkowski : *Multicriterion Discrete Optimization of Space Trusses with Serviceability Constraints*

M. Pyrz : *Symbolic Computations Approach in Controlled Enumeration Methods Applied to Discrete Optimization*

A. T. Janczura : *General P-$\triangle$ Method in Discrete Optimization of Frames*

Sesion 7 : **Applications of genetic algorithms and neuro networks - Round table discussion**
Chairman : G.I.N. Rozvany

Introductory remarks by : P. Hajela and Ph. Trompette

Contents

Optimal Damping of Beams and Plates by Genetic Algorithms 1
Ph. Trompette, J. L. Marcelin, C. Schmeding

Optimization by the Voting Method of Structures Formed
of Planar Constitutive Parts 12
J. Mottl

Neuro-Optimizer, Its Application to Discrete Structural
Optimization 22
M. Kishi, T. Kodera, Y. Iwao, R. Hosoda

Genetic Algorithms in Topological Design of Grillage
Structures 30
P. Hajela, E. Lee

Optimization of a Linear Objective Function with Logical
Constraints 40
J. Mączyński

On Optimal Structural Segmentation Problem 47
K. Dems, Z. Mróz

Application of Discrete Optimization Techniques to Optimal
Composite Structures 61
E. Schäfer, J. Geilen, H. A. Eschenauer

A Random Search Approach to Multicriterion Discrete
Optimization 71
A. Osyczka, J. Montusiewicz

Optimal Arrangement of Ribs in Rib Reinforced Plates
and Shells 80
A. O. Rasskazov, A. S. Dekhtjar

Application of the Expert System for Discrete Optimization
of Space Truss 88
J. Niczyj, W. Paczkowski

Discrete Structural Optimization Based on Multipoint Explicit
Approximations 98
V. V. Toropov, V. L. Markin, H. Carlsen

Optimization of Perturbation Parameters for Forced Vibration
Stress Levels of Turbomachinery Blade Assemblies 108
E. P. Petrov

New Discretized Optimality Criteria Methods - State
of the Art. 118
G. I. N. Rozvany, M. Zhou

Discretized Methods for Topology Optimization 135
G. I. N. Rozvany, M. Zhou, T. Birker, T. Lewiński

Homology Design of Flexible Structure by the Finite
Element Method 148
N. Yoshikawa, S. Nakagiri

Optimization of Structure and Development of Production
System 158
S. Shevchenko

Support Number and Allocation for Optimum Structure 168
W. Gutkowski, J. Bauer, Z. Iwanow

Sensor Allocation for State and Parameter Estimation
of Distributed Systems 178
J. Korbicz, D. Uciński

Optimal Location of Piezoelectric Actuators 190
J. Holnicki-Szulc, F. Lopez-Almansa, A. Mackiewicz

Optimal Discrete Design of Elastoplastic Structures 200
A. Čižas, S. Stupak

Multicriterion Discrete Optimization of Space Trusses with
Serviceability Constraints 209
S. Jendo, W. M. Paczkowski

Symbolic Computations Approach in Controlled Enumeration
Methods Applied to Discrete Optimization 221
M. Pyrz

General P-Δ Method in Discrete Optimization
of Frames 228
A. T. Janczura

A More General Optimization Problem for Uniquely Decodable
Codes 239
A. A. Belal

Optimal Damping of Beams and Plates by Genetic Algorithms

Ph. Trompette, J.L. Marcelin, C. Schmeding

Laboratoire **3S**   URA CNRS 1511   BP 53 X 38041 GRENOBLE   FRANCE

Abstract. This paper is concerned with the optimal damping of beams or plates partially covered by viscoelastic constrained layers. The objective function is to maximize the modal damping factor of one (or several) mode(s). During the optimization process, the selected mode shapes are those of the undamped associated structure. A convenient finite element model for the beam or the plate dynamic analysis is used. The possible dimensions and locations of the viscoelastic layers are determined by the use of a genetic algorithm.

Keywords. Optimal Damping, Viscoelastic, Beams and Plates, Genetic Algorithms.

1 Introduction

Structural vibration control is a major design problem for a variety of structures. This control may be approached in several ways as active attenuators, structural damping, etc... In most cases the objective of the designers is to minimize the vibration amplitudes on a wide range of frequencies to prevent damage by fatigue. For such a purpose, the use of viscoelastic constrained layers may be an interesting solution compatible with technological constraints. In |3|, the optimal modal damping of beams partially covered by viscoelastic layers has been studied ; the minimization was performed using a conventional non-linear programming algorithm.

For plates, a very few number of papers have been previously published. |4| and |5| present theoretical and experimental studies of plates with complete covering, and |6| with partial one, but without an optimization step. For this kind of optimization problem, because of the difficulty to define a convenient design variable set able to initiate an appropriate initial design point, the usual non-linear programming methods are not well suited.

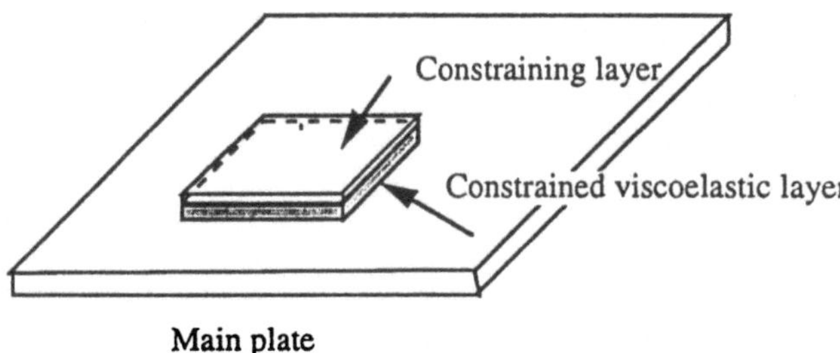

Figure 1 — Plate damped by viscoelastic constrained material

In this paper, both partial covering and optimization are considered. For such a purpose, special plate finite elements are defined which represent the behavior of the sandwich parts ; they are described in the next section. On the other hand, a genetic algorithm is used to optimize the dimensions and the locations of the viscoelastic layers. As in |1| or in |6| the main conclusion in achieving an optimal damping consists of the realization that it is useless to cover the whole structure ; in |6| , both theory and experiment show that for stiff viscoelastic layers, the loss factor is greater with a partial coverage than with a full one.

2 The finite element model

The dynamical behavior -i.e.- frequencies and mode shapes- of partially covered plates, is calculated from a modal finite element model. The homogeneous parts of the plate are discretized by conventional C_1 F.E. and the heterogeneous or sandwich ones by specific F.E. designed to represent accurately the viscoelastic core shear damping effect; such elements were proposed previously in |2| and |3| for beams and are extended here to plates.

2.1 The finite elements

2.1.1 Homogeneous plate F.E.

It is an usual Kirchhoff plate bending element with four nodes. The degrees of freedom are w, $\partial w/\partial x, \partial w/\partial y$ where w is the transverse plate displacement. The interpolation function w is,

$$w = a_1 + a_2x + a_3y + a_4xy + a_5x^2 + a_6y^2 + a_7x^3 + a_8y^3 + a_9x^2y + a_{10}xy^2 + a_{11}x^3y + a_{12}xy^3 \tag{1}$$

In the following applications the in-plane displacements of the main plate are neglected.

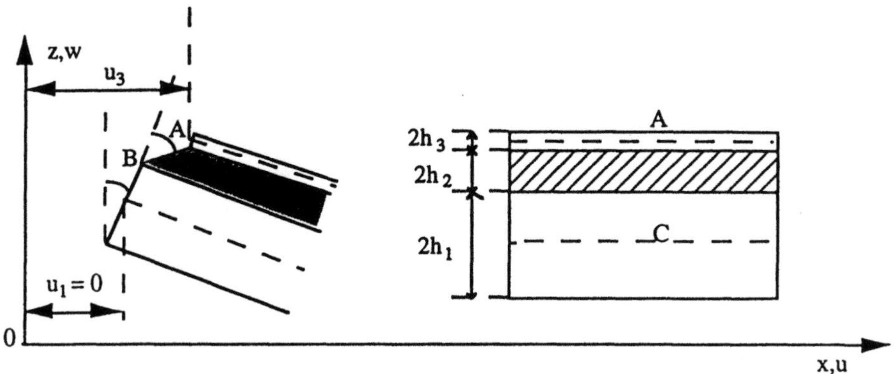

Figure 2 - Displacements of the layers

2.1.2 Heteregeneous F.E.

It is built on the following assumptions :
-the transverse plate displacement w is independent of the transverse coordinate z
-the viscoelastic layer bending stiffness is negligible and the strains are only shear strains
-the longitudinal displacements of the different parts of the plate (Figure 2) are given by

$$u_A = u_3 + h_3(\partial w/\partial x), \quad u_B = - h_1(\partial w/\partial x) \tag{2}$$

and
$$v_A = v_3 + h_3(\partial w/\partial y), \quad v_B = - h_1(\partial w/\partial y)$$

- the viscoelastic core material is characterized by a complex shear modulus G which can be frequency and temperature dependent.
- the thicknesses $h_3 \ll h_1$ and $h_2 < h_1$
-the shear strain energy of the elastic layers is neglected.
-the longitudinal displacements u and v of the plate medium surface are set to be 0.
-the constrained viscoelastic layer is only subjected to shear strains γ_{xz} and γ_{yz}.

At each node the degrees of freedom are w, $\partial w/\partial x, \partial w/\partial y$ and the longitudinal displacements u and v of the constraining layer. The interpolation function for w is the same as that for the homogeneous element. The interpolation functions for u and v are those of the bilinear plane element.

The element stiffness matrix is the sum of the 3 contributions : bending, in-plane traction of the constraining layer, and in-plane shear of the constrained viscoelastic layer.

4

The first two are usual, the last is obtained from the following in-plane shear strain energy :

$$E = \frac{1}{2} \int_{(A)} \int_{h_1}^{h_1+2h_2} (G\gamma_{xz}^2 + G\gamma_{yz}^2) \, dz \, dA \tag{3}$$

with :
$$\gamma_{xz} = \frac{\partial u_2}{\partial z} + \frac{\partial w}{\partial x} \quad \text{and} \quad \frac{\partial u_2}{\partial z} = \frac{u_A - u_B}{2h_2}$$

so :
$$\gamma_{xz} = \frac{1}{2h_2}(u_3 + d\frac{\partial w}{\partial x}) \quad \text{with} \quad d = h_1 + 2h_2 + h_3 \tag{4}$$

in the same manner :
$$\gamma_{yz} = \frac{1}{2h_2}(v_3 + d\frac{\partial w}{\partial y}) \tag{5}$$

2.2 Element mass matrix

A consistent mass matrix M is used which takes account of the bending contribution and of the in-plane effects of the different parts, i.e. the calculated kinetic energy is:

$$T = \frac{1}{2} \int_1 \rho_1 \dot{w}^2 d\tau + \frac{1}{2} \int_2 \rho_2 (\dot{w}^2 + \dot{u}_2^2 + \dot{v}_2^2) d\tau + \frac{1}{2} \int_3 \rho_3 (\dot{w}^2 + \dot{u}_3^2 + \dot{v}_3^2) d\tau \tag{6}$$

2.3 Assembly

Due to the viscoelasticity of the core material the global dynamic equilibrium equations are simple only if the displacements are periodic : $x = x_0 e^{j\omega t}$. In this case, the viscoelastic modulus is written $G = G_0 (1 + j\beta)$ and the corresponding stiffness matrix is

$$|K_v| = |K_{vr}| + j|K_{vi}| \tag{7}$$

The dynamic behavior of the whole plate (or beam) is given by the matrix equation

$$(-|M|\omega^2 + |K_1| + j|K_2|)\{u\} = \{F\} \tag{8}$$

in which the imaginary part K_2 of the total stiffness matrix is equal to K_{vi}.

2.4 Modal response

Frequencies and mode shapes of the undamped associated structure can be considered as a good and simple modal basis to be used for predicting the dynamic behavior of the damped structure [9].

ω_i and $\emptyset_i$ $i = 1, n$ are the undamped frequencies and corresponding mode shapes obtained from the matrix equation :

$$(-|M|\omega^2 + |K_1|)\{x\} = \{0\} \tag{9}$$

Performing the usual transformation $\{x\} = |\Phi| \{q\}$, and premultiplying by $|\Phi|^T$, equation (10) is obtained for free vibrations

$$(-|\mu|\omega^2 + |\kappa_1| + j|\kappa_2|)\{q\} = \{0\} \tag{10}$$

Because of the orthogonality of the modes, $|\mu|$ and $|\kappa_1|$ are diagonal matrices, but not $|\kappa_2|$. Generally for beams and plates the frequencies are well separated, so the full matrix can be considered as diagonal dominant. In these conditions the modal system (10) is the sum of n uncoupled equations. It follows that in a modal response, a good approximation of the structural loss factor may be easily calculated from each equation of (10)[1].

3 The Optimization Method

The objective function to maximize is the modal damping factor of a single mode i or a linear combination of modal damping factors for several i.

$$\text{Max } \eta = SE_{vi}/SE_i \quad \text{or} \quad \Sigma_i \, a_i SE_{vi}/SE_i \tag{11}$$

In |11| SE_{vi} is the elastic strain energy stored in the viscoelastic material when the structure vibrates on its ith undamped mode, and SE_i is the corresponding elastic strain energy of the entire composite structure, a_i is a weighting coefficient. As all the matrices are diagonal and the mode shapes are non-variable during the optimization process, the calculation of the damping factors is obviously straightforward.

$$\eta = \emptyset_i^T |K_2|\emptyset_i/\emptyset_i^T |K_1|\emptyset_i = \kappa_{2i}/\mu_i\omega_i^2 \quad \text{or} \quad \Sigma_i \, a_i \, ...$$

By using conventional optimization procedures (non linear programming) it is impossible to solve this optimization problem because of the great difficulty in determining a convenient starting point in the design space and also an efficient optimization strategy. This is the reason why here, a genetic algorithm is used. Genetic algorithms (GAs) are random search procedures based on natural selection (|7| and |8|).

---

[1] If the shear modulus of the viscoelastic core is frequency and temperature dependant, an iterative method to calculate the resonance frequencies and mode shapes must be used |2|.

A simple genetic algorithm consists of three operators: reproduction, crossover, and mutation.They are particulary well suited to represent simply the dimensions and the locations of the viscoelastic layers.

In the present paper, heterogeneous plate elements are coded by 1 and homogeneous plate elements by 0, so a design point - a chromosome- is a n binary number in which n is the number of finite elements. The genetic algorithm used is as described in |8|. There are no particular limitations and constraints.

4 Results

4.1 Results with a fixed mesh

At first, tests are conducted on a beam (or a narrow plate) defined in |1|. The purpose is to compare the results given by the present work (using a GA) and those obtained in |1| by both experimental and theoretical methods. The dimensions of the plate are the following : Length L = 508 mm, width b = 50.8 mm, h_1 = 3.175 mm, h_2 = 1.27 mm, h_3 = 0.787 mm.

The viscoelastic material shear modulus G_0 is equal to 16.3 MPa, the loss factor $\beta = 1$ (in this example both are supposed constant) and the specific gravity of the viscoelastic material is 2500 kg/m^3 . The elastic material is steel. Nine plate elements of length 28.22 mm are used to model half of the plate.

The optimization is conducted on the first clamped-clamped mode. Only half of the beam is studied ; the length of the chromosome is 9, the population size is 10, the maximum number of generation is 8, the crossover probability is 6.0E-01 and the mutation probability is 3.33E-02. For this first mode, the results are quantitatively good when compared with those of |1|. The final size and location of the viscoelastic layers are given in Fig. 3. The optimal structure (genetic solution) is made of 3 heterogeneous plate elements followed by 1 homogeneous plate element, followed by 10 heterogeneous elements followed by 1 homogeneous element, followed by 3 heterogeneous elements. It is coded by 111011111 for half of the plate. The damping factor, as defined by (4), increases from 0.16705 (for the whole plate covered) to 0.20205. The optimal continuous solution of |1| is also given in figure 3 (damping factor 0.17044). In this example the genetic solution is better than the continuous one.

Then, the optimization is conducted for the first free-free mode. For this first mode, the results are quantitatively good as compared with |1|. All the results are given in figure 4. It is found that the continuous solution is the same as the genetic solution.

The damping factor, as defined by (4), increases from 0.22240 (for the whole plate covered) to 0.28400. During the genetic procedure another interesting solution occurs for which the damping factor is 0.28338. This solution is coded by 010111111 for the half plate.

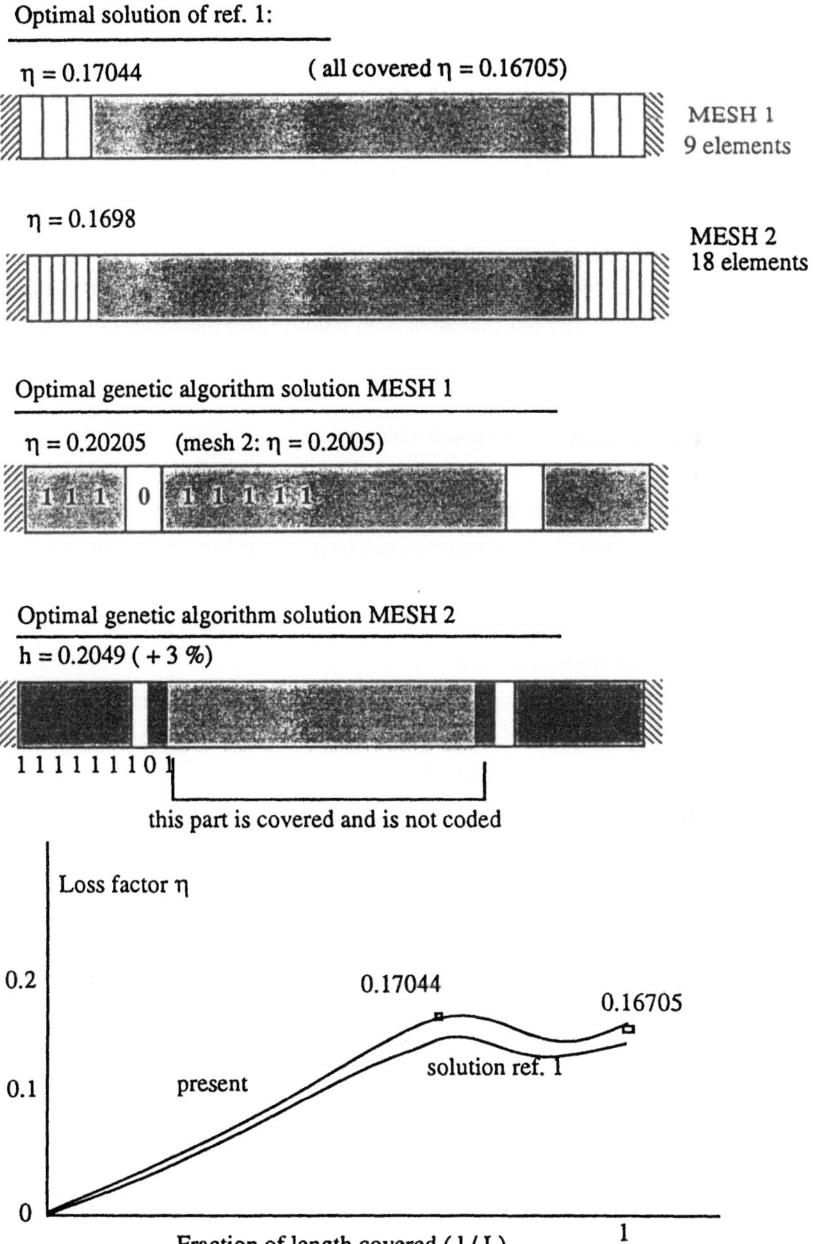

Figure 3 — Optimal covering for the first clamped-clamped bending mode

8

The second example is concerned with a square plate dimensions of which are : length L = 6 m, width b = 6 m, h_1 = 0.003 m, h_2 = 0.0015m, h_3 = 0.0002 m.

The viscoelastic material shear modulus G_0 is equal to 126.7 MPa, the loss factor β =1 (both are constant) and the specific gravity of the viscoelastic material is 2500 kg/m^3 . The elastic material is aluminium.

Nine plate elements are used to model a quarter of the plate. The optimization is conducted for the first clamped-clamped mode. Only a quarter of the plate is studied ; the length of the chromosome is 9, the population size is 10, the maximum number of generation is 10, the crossover probability is 8.0E-01 and the mutation probability is 6.0E-02. All the results of this case are given in figure 5 .

The final structure (genetic solution) is coded by 101010101 for the quarter plate. The damping factor, as defined by (4), increases from 0.00413 (for the whole covered plate) to 0.00688 for the optimal solution.

4.2 Results with a mesh refinement strategy

In the preceeding results it can be pointed out that some blocks of binary digits do not change in the chromosoms. This observation can be associated to a FE mesh refinement strategy. These blocks are not coded and optimization is performed only on the variable parts of the chromosomes. In this manner, the dimensions of the chromosomes become less important when the mesh is refined.

With this mesh refinement strategy, all the preceding examples were studied again (Figures 3, 4 and 5). For the beam, results (mesh 2) are improved to a proportion of 3 at 4 %, and for the plate to a proportion of 28%.

5 Conclusion

The optimal modal damping of beams or plates, partially covered by constrained viscoelastic layers, is calculated using a genetic algorithm. The size and locations of the layers are the design variables. It is demonstrated that the genetic algorithm is an efficient procedure for solving a problem for which the classical non linear programming methods are unadapted.

Optimal solution of ref.1 and genetic optimal solution:

$\eta = 0.28400$ (all covered = 0.2224)

Secondary genetic algorithm solution MESH 1

$\eta = 0.28338$ (mesh 2 : $\eta = 0.28424$)

Optimal genetic algorithm solution MESH 2

$\eta = 0.2970 \, (+ 3.9 \, \%)$

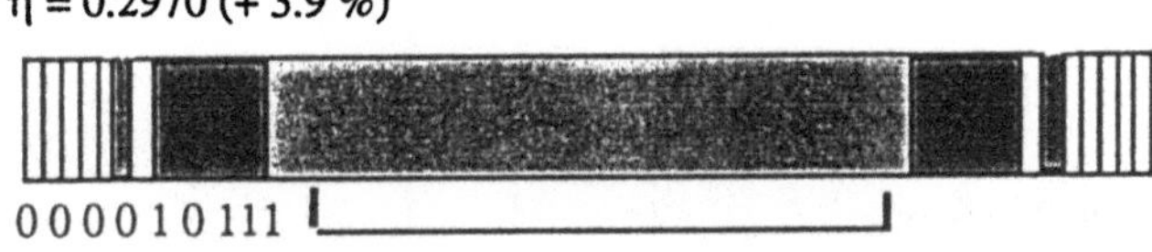

0 0 0 0 1 0 111

this part is covered and is not coded

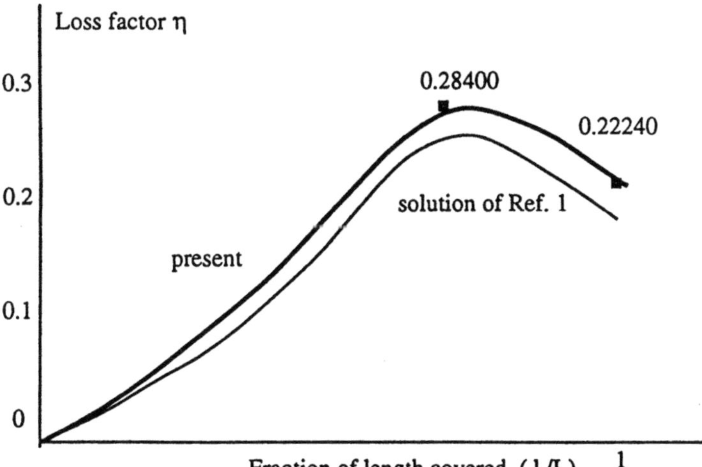

Figure 4 — Optimal covering for the free-free first bending mode

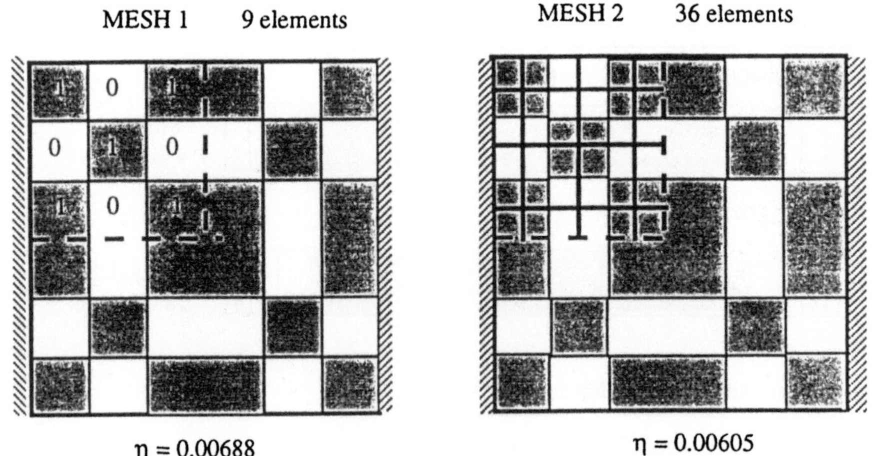

MESH 2 : quarter of the plate

FIGURE 5 : clamped-clamped plate

Figure 5 — Optimal damping and mesh refinement

References

|1| D. S. NOKES and F. C. NELSON, "Constrained layer damping with partial coverage", *Shock Vib. Bull.* **38**, pp.5-12,1968.

|2| P. TROMPETTE, D. BOILLOT and M.-A. RAVANEL, " The effect of boundary conditions on the vibration of a viscoelastically damped cantilever beam", *J. Sound Vib.* **60** (3), pp.345-350,1978.

|3| J.-L. MARCELIN, Ph. TROMPETTE, and A. SMATI, "Optimal constrained layer damping with partial coverage", *Finite Elements in Analysis and Design* **12**, pp. 273-280,1992.

|4| Y.P. LU, J.W. KILLIAN and G.C. EVERSTINE, " Vibrations of three layered damped sandwich plate composites", *J. Sound Vib.* **64**(1), pp.63-71,1979.

|5| J.-F. HE and B.-A. MA, " Analysis of flexural vibration of viscoelastically damped sanwich plates", *J. Sound Vib.* **126**(1), pp.37-47,1988.

|6| Y. JULLIEN and T. TAKAGAMI, "Vibration de poutres et de plaques recouvertes en partie de matériaux amortissants",*Acustica* **47**, pp.304-313, 1981.

|7| P. HAJELA, "Genetic Search - An approach to the nonconvex optimization problem", *AIAA Journal* **28**(7) pp. 1205-1210 ,1990.

|8| D.E. GOLDBERG, <u>Genetic Algorithms in Search, Optimization, and Machine Learning</u>, Addison-Wesley, Reading, MA,1989.

(9) C.J. WILSON, P. CARNEVALI, R.B. MORRIS and Y. TSUJI, "Viscoelastic Damping Calculations Using a p-Type Finite Element Code", *Transactions of the ASME* **59** pp. 696-699 ,1992.

Optimization by the Voting Method of Structures Formed of Planar Constitutive Parts

J. Mottl,
Tererova 6, 61200 Brno, CZECH REPUBLIC

1. Introduction

Numerous approaches are known to exist in structural optimization. To the oldest ones belongs the *full stress method*, it consists in successive steps of calculations and corrections. Later methods originated which use methods of linear and nonlinear programming. Often methods of linear approximation are used and these transform nonlinear problem into a sequence of linear programming solutions. No neglect of either methods or their authors is intended in this paper but a few only may be mentioned.

Most treatments introduced work mainly with continuous parameters (thicknesses of walls, profiles etc.). However, normalized dimensions are used when manufacturing plates and profiles. Therefore the optimization as a problem with discrete parameters is closer to reality. The methods of integer programing are the methods used in this field. The following contribution is concerned with the discrete optimization of space structures formed of planar parts.

2. The problem formulation

Let us consider a space structure formed of planar parts (Fig.1). This structure is partitioned into $I = 1, N2C$ "bodies I" (in Fig.1 $N2C = 8$). Each body I consists then of $JME(I)$ triangular elements e.(In Fig.1 is $JME(I) = 2, I = 1, N2C$). The first N2 of these bodies are then the "bodies"the thickness b of which will be changed during the solving (in the remaining bodies the thicknesses will be maintained). Each body $I = 1, N2$ can select its own thickness b from the sample of variants, given by the list of input

$$B2(I, IZ), \ I = 1, N2; \quad IZ = 1, IZM.$$

Within the considered structure we shall require that the maximal reduced stress in each body should be less than the admissible stress σ_D and that simultaneously the structure should be of minimum weight. This can be expressed by following relationships:

$$G(I1) = \sigma_r^{max}(I = I1.e) - \sigma_D \leq 0 \quad I1 = 1, N2C \tag{1}$$

$$f = G(N2C + 1) = \sum_{=1}^{N2} B2(I, IZ)BA(I) \rightarrow \min. \tag{2}$$

Here the $G(I1)$ denotes the $I1$ constraint (the $I1$ index is reserved for the constraint counter. Because this condition is written for each body $I1, I1 = I$ also holds, the notation I being reserved for body counter). $\sigma_r^{max}(I1, e)$ means the reduced stress in the element e of the body I where it reaches the maximal value. σ_D is the admissible stress. $N2C$ is the total number of "bodies", $N2$ is the number of such bodies where the thickness b_i is selected from the list $B2(I, IZ)$; $BA(I)$, $I = I, N2$ are the plane areas of individual bodies I.

The (1),(2) formulation represents a problem of nonlinear programming with the system of constraint conditions (1) and the objective function (2). The goal is to find such variants IZ of thicknesses of individual bodies I that the weight of the structure attains a minimum.

3. The determination of stress in the wall

A plane state of stress is assumed to occur in the "bodies" constituting the structure. We shall consider a finite element e in a body I (triangular in form with nodal points i, j, k and with x, y coordinate system fixed in it - see Fig.2). Then, according to (9) we can find the direction cosines (of the local coordinate system x, y of the element e consided with respect to the global coordinate system X, Y, Z) in the form S^e (with dimensions 6×9), the matrix B (with dimensions 3×9) follows from the coordinates of nodal points i, j, k in local coordinates x, y of the element e.

Then the rigidity matrix of the element e in local coordinate system (with dimensions 6×6) is:

$$k^e = b_1 AB^{et} DB^e \tag{3}$$

where A is the of the element area, b_1 is its thickness (e is a part of body I), D (3×3) is the matrix of physical constants.

Then the rigidity matrix of the element e in the global coordinate system is (9×9)

$$k_G^e = \lambda^{et} k^e \lambda^e \tag{4}$$

The rigidity matrix K of the whole structure is assembled by summation over all elements k_G^e. Then the system of equations for the displacements of the construction nodal points Q has the form

$$KQ = F \tag{5}$$

Here F is the loading vector. Taking support conditions into account we obtain from Eq.(5) the vector of the displacements of the nodal points in the global coordinate system.

14

$$Q = \{u_1, v_1, w_1, ..., u_{IBM} v_{IBM} w_{IBM}\}$$

where IBM is the number of the nodal points of the construction. The dimension of Q is $1 \times N$, $N = 3 \times IBM$.

Let us return to the element e considered in body I. From Q can be obtained $Q^e = \{u_i, v_i, w_i, u_j, v_j, w_j, u_k, v_k, w_k,\}$ i.e. the displacements of nodal points i, j, k of element e in global coordinate system X, Y, Z. From them can be obtained $q^e = \{\bar{u}_i, \bar{v}_i, \bar{w}_i, \bar{u}_j, \bar{v}_j, \bar{w}_j, \bar{u}_k, \bar{v}_k, \bar{w}_k,\}$ i.e. the displacements of the nodal points i, j, k of the element e in its local coordinate system x, y in form:

$$q_e = \lambda^e Q^e \tag{6}$$

Then the vector of stresses in the element e considered has the form

$$\sigma^e = \left\{ \begin{array}{c} \sigma_{xx} \\ \sigma_{yy} \\ \sigma_{xy} \end{array} \right\} = D^e B^e q^e = D^e B^e \lambda^e Q^e = A^e_m Q^e \tag{7}$$

or the σ^e in the element e is the function of displacements of its nodal points in global coordinate system Q^e. A_m has dimensions 3×6.

Then using components of σ^e we can find the value of reduced stress in element e.

$$\sigma^e_r = \frac{1}{\sqrt{2}} \sqrt{\sigma^2_{xx} - \sigma_{xx}\sigma_{yy} + \sigma^2_{yy} + 3\sigma^2_{xy}} \tag{8}$$

4.1 The determination of the sensitivity matrix

In the following considerations we shall need the characteristics, which would define with what intensity will an individual constraint $G(I1)$ in (1) change when the thicknesses b_i of individual bodies I of the structure are modified. Let us denote:

$$G = \left| \begin{array}{c} G(I1 = 1) \\ \vdots \\ G(I1 = N2C) \end{array} \right| \qquad b = \left| \begin{array}{c} b_{I=1} \\ \vdots \\ b_{I=N2} \end{array} \right| \qquad Q = \left| \begin{array}{c} \delta_1 \\ \vdots \\ \delta_N \end{array} \right| \tag{9}$$

Then the searched for characteristic can be found from the relation

$$\frac{\partial G}{\partial b} = \frac{\partial G}{\partial Q} \frac{\partial Q}{\partial b} \tag{10}$$

where

$$\frac{\partial G}{\partial b} = \left| \begin{array}{ccc} \frac{\partial G(I1=1)}{\partial b_{I=1}} & \cdots & \frac{\partial G(I1=1)}{\partial b_{I=N2}} \\ \vdots & & \vdots \\ \frac{\partial G(I1=N2C)}{\partial b_{I=1}} & \cdots & \frac{\partial G(I1=N2C)}{\partial b_{I=N2}} \end{array} \right| \tag{11}$$

$$\frac{\partial G}{\partial Q} = \begin{vmatrix} \frac{\partial G(I1=1)}{\partial \delta_1} & \cdots & \frac{\partial G(I1=1)}{\partial \delta_N} \\ \vdots & & \vdots \\ \frac{\partial G(I1=N2C)}{\partial \delta_1} & \cdots & \frac{\partial G(I1=N2C)}{\partial \delta_N} \end{vmatrix} \tag{12}$$

$$\frac{\partial G}{\partial b} = \begin{vmatrix} \frac{\partial \delta_1}{\partial b_{I=1}} & \cdots & \frac{\partial \delta_1}{\partial b_{I=N2}} \\ \vdots & & \vdots \\ \frac{\partial \delta_N}{\partial b_{I=1}} & \cdots & \frac{\partial \delta_N}{\partial b_{I=N2}} \end{vmatrix} \tag{13}$$

4.2 Determination of $\frac{\partial G}{\partial Q}$

The row $I1$ of the matrix $\frac{\partial G}{\partial Q}$ according to (12) is $\frac{\partial G(I1)}{\partial Q}$. Because of (1)

$$\frac{\partial G(I1)}{\partial Q} = \frac{\partial \sigma_r^{max}(I = I1, e)}{\partial Q} = \frac{\partial \sigma_r^{max}(I = I1, e)}{\partial Q^e} \tag{14}$$

σ_r is defined according to (8) and can be written in matrix form:

$$2\sigma_r^2 = [\sigma_{xx}\sigma_{yy}\sigma_{xy}][\sigma_{xx} - \sigma_{xx}\sigma_{yy}3\sigma_{xy}]^T \tag{15}$$

The matrix A_m^e according to (7) has the following components

$$A_m^e = \begin{bmatrix} C_{xx} \\ C_{yy} \\ C_{xy} \end{bmatrix} \tag{16}$$

Further we shall define the matrices (each with dimensions 4×6)

$$A_1^e = \begin{bmatrix} C_{xx} \\ C_{xx} \\ C_{yy} \\ C_{xy} \end{bmatrix} \qquad A_2^e = \begin{bmatrix} C_{xx} \\ -C_{yy} \\ C_{yy} \\ 3C_{xy} \end{bmatrix} \tag{17}$$

Then (15) can be written in the form

$$L = 2\sigma_r^2 = (A_1^e Q^e)^T A_2^e Q^e \tag{18}$$

Then for $\frac{\partial L}{\partial Q^e}$ (where Q^e, $l = 1,9$ are the components of the displacements of nodal points of element e in global coordinate system) holds:

$$\frac{\partial L}{\partial Q_l^e} = (A_{1,l}^{eT} A_2^e + A_{2,l}^{eT} A_1^e)Q^e, \quad l = 1,9 \tag{19}$$

In this expression $A_{1,e}^e$ ($A_{2,e}^e$ respectively) means the l-th column of the A_1^e (A_2^e respectively) matrix.

Using (19) the Il-th row can be determined in $I = 1, 9$ column in matrix $\frac{\partial G(I1)}{\partial Q}$. If it is performed for all $I1 = 1, N2C$ we get the $\frac{\partial G}{\partial Q}$ matrix.

4.3 The determination of $\frac{\partial Q}{\partial b}$

For $\frac{\partial Q}{\partial b}$ with respect to (5) is valid:

$$\frac{\partial Q}{\partial b} = \frac{\partial}{\partial b}(K^{-1}F) = \frac{\partial K^{-1}}{\partial b}F \tag{20}$$

The difficulty of finding $\frac{\partial K^{-1}}{\partial b}$ can be obviated using the relation

$$KK^{-1} = 1 \tag{21}$$

Therefore

$$K\frac{\partial K^{-1}}{\partial b} + \frac{\partial K}{\partial b}K^{-1} = 0 \tag{22}$$

From here we can find $\frac{\partial K^{-1}}{\partial b}$ and by its substitution into (20) we obtain

$$\frac{\partial Q}{\partial b} = -K^{-1}\frac{\partial K}{\partial b}K^{-1}F = -K^{-1}\frac{\partial K}{\partial b}Q =$$

$$= -K^{-1}|\frac{\partial K}{\partial b_{I=1}} \frac{\partial K}{\partial b_{I=2}} \cdots \frac{\partial K}{\partial b_{I=N2}}|Q \tag{23}$$

For example $\frac{\partial K}{\partial b_{I=1}}$ in (23) is the matrix in which all components are zero with exception of those which correspond to those finite elements e which form the body $I = 1$ (and $b_{I=1}$ equal 1) etc.

5. The steps of the proposed optimization

1. The solving has an iterative character. We shall begin with iteration $IT = 1$ and put $IMC = N2C$.

2. We choose the arbitrary starting solution $IX(I), I = 1, N2$ - which means a certain selection of IZ variants (from possibles) in bodies I. For example the meaning of $IX(I) = (3\ 2\ 6\ ...)$: in $I = 1$ is its variant $IZ = 3$, in body $I = 2$ is its variant $IZ = 2$ etc.

In this way we know the concrete thicknesses $b. = B2(I, IX(I))$ in individual bodies I and we can find the values of $k^e \rightarrow k^e_G \rightarrow K \rightarrow Q \rightarrow Q^e \rightarrow \sigma^e \rightarrow \sigma^e_r$ in all finite elements e of the construction, step by step, according to Chapter 3.

3. In each body $I = 1, N2C$ the element e with maximal σ^e_r will be found and in this way according to (1) we can find the values of limiting conditions $G(I1), I1 = 1, N2C$ in the iteration considered.

4. Now the test of feasibility of solution $IX(I)$ follows. The criterion is that all constraints must be negative i.e.

$$G(I1) \leq 0; \quad I = 1, IMC \tag{24}$$

If (24) is not fulfilled for all $I1$ the process goes to point No.5. If (24) is fulfilled then the solution $IX(I), I = 1, N2$ is printed together with the message FEASIBLE DOMAIN. In the following we put $IMC = N2C + 1$ and from the objective function (2) a supplementary constraint is created in the form:

$$G(N2C + 1) = G(N2C + 1) + [-G(N2C + 1) : IZ = IX(I) + DELT] \tag{25}$$

where DELT is a chosen positive number. In this way to the system $G(I1) \leq 0$ for $I1 = 1, N2C$ is added a new constraint $G(N2C + 1) = DELT > 0$ which compels the searching algorithm to search a new solution $IX(I)$. This new solution should have the property that by preservation of feasibility (i.e. negative values) of constraints $G(I1)$ introduced also the $G(N2C+1)$ will be negative (feasible), or the f with smaller value (weight of the construction). See [5].

5. The own "voting process": Each "body I" which functions in certain limitation $I1$ will "vote" for its own individual variants IZ according to the rule:

$$HL(I1, I, IZ) = |\bar{G}(I1)\frac{\partial G(I1)}{\partial b_1}|(\overset{+}{-} SZ(I, IZ)) \tag{26}$$

In this expression $HL(I1, I, IZ)$ is the value of the vote which gives the body I in limitation $I1$ to its certain variant IZ. SZ is a chosen function according to the Fig.2. The $+SZ$ has for increasing coordinates IZ the values mostly positives (beginning with the coordinate $IZ = IX(I)$). For IZ lesser than $IX(I)$ the values of $+SZ$ are negative. The $-SZ$ has the distribution introduced inverse (see Fig.2). Further the parameter ID is introduced, which defines how many of the first elements in SZ (beginning from $IX(I)$) will be used in voting. The problem how to choose the SZ we shall discuss later.

The rule for the choice of $+SZ$ or $-SZ$ in the relation (26) is the following one

$$\frac{\partial G(I1)}{\partial b_1} > 0 \rightarrow -SZ(I, IZ)$$

$$\frac{\partial G(I1)}{\partial b_1} < 0 \rightarrow +SZ(I, IZ) \tag{27}$$

or $G(I1)$ in (26) is taken

$$G(I1) > 0 \rightarrow \bar{G}(I1) = G(I1)$$

$$G(I1) < 0 \rightarrow \bar{G}(I1) = C20(G(I1) > 0)_{min}/(1 + |G(I1)|) \tag{28}$$

The relation (26) can be interpreted as follows: if $\frac{\partial G(I1)}{\partial b_1} > 0$ is valid then the increase of b_1, would lead in this case to a increase of $G(I1)$ and therefore to the increase of the unfulfilment of this constraint. To prevent this situation we will choose - in the considered body I - its variants IZ with lesser variants b_1 - and this is identical with the choice of $-SZ(I, IZ)$ for voting.

If $\frac{\partial G(I1)}{\partial b_1} < 0$ is valid, then an opposite situation is created and we prefer rather the variants IZ with greater b_1 it means that we use $+SZ(I, IZ)$ for voting in (26).

For $G(I1)$ in (26) we take $G(I1)$ in case if $G(I1) > 0$ and if $G(I1) < 0$ the relation (28b) (independently if $\frac{\partial G(I1)}{\partial b_1}$ is positive or negative). The constant $C20$ in (28b) is given some attention below. The goal in the "voting" so defined is that such variants IZ of individual bodies I are preferred, which can contribute to the situation that the values of constraints $G(I1)$ will be lowered - therefore become closer to negative (feasible) domain (solution).

This voting has in fact to be performed only in case of $G(I1) > 0$ constraints - i.e. in case of unfulfilment of certain constraint. But according to (28b) the voting is performed also in case of $G(I1) < 0$, i.e. in case of fulfilled constraint, but with a lower weight than the unfulfilled constraints. According to (28b) the weight decreases with the increase of the fulfillment i.e. $G(I1) < 0$. The reason is to prevent the constraint from turing positive in the next iteration.

6. The next step is obvious. We have to find

$$HLV(I, IZ) = \sum_{I1=1}^{IMC} HL(I1, I, IZ) \tag{29}$$

This expression represents the result voting evaluation of individual variants IZ of the body I from the view point of all limitations $I1 = 1, IMC$. See Fig.3. Then the new solution will be evidently:

$$IX(I) = IZ : HLV^{max}(I, IZ), \quad I = 1, N2 \tag{30}$$

i.e. the new variant IZ of body I is this IZ for which the vote $HLV(I, IZ)$ is maximal.

7. The next iteration $IT - -T + 1$ follows and with the new $IX(I)$ we go into the point 2 of the procedure. The last one from the sequence of solutions obtained in this way $IX(I)$ with the message FEASIBLE DOMAIN is then the required extremum. The process terminates if during of ITM iterations after the last F.D. the new F.D. is not attained.

5.1 Convergence of the process

The practical realization of the process showed, that for finding the next feasible solution $IX(I)$ (with F.D.) several iterations are necessary and the values (and signs) of individual $G(I1)$ oscillate during iterations. This process (reaching of F.D.) is positively influenced by the situation, when "the votes for the change" - i.e. generated from $G(I1) > 0$ constraints - are in approximate equilibrium with votes which are against the change of $IX(I)$ $(G(I1) < 0)$.

This requirement can approximately be formulated by relation for the constant $C20$ determination in (28b) in form

$$BK \sum_{G(I1)>0} \sum_{I=1}^{N2} |G(I1)\frac{\partial G(I1)}{\partial b_1}| = C20(G(I1) > 0)_{min} \cdot$$

$$\cdot \sum_{G(I1)<0} \sum_{I=1}^{N2} |G(I1)\frac{\partial G(I1)}{\partial b_1}/(1 + |G(I1)I|)| \tag{31}$$

Here BK is the chosen weight $(1-1.2)$ for the relation of the votes for and against the change.

5.2 Illustration example

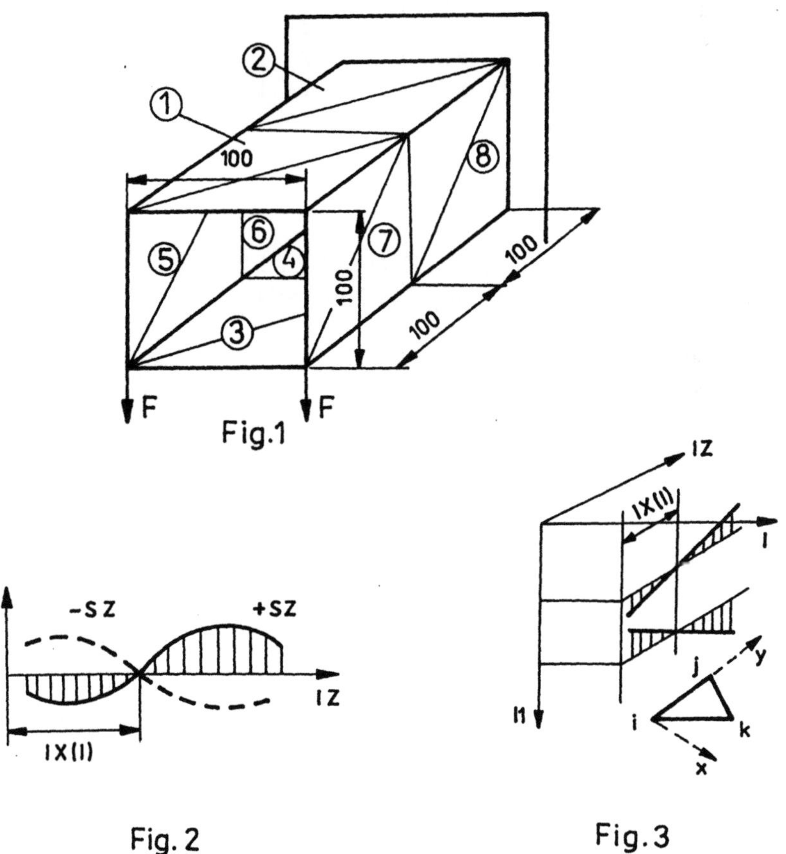

20

As an illustration example a simple example according to Fig.1 was chosen with $F = 10^3 \ N$ $\sigma_D = 3.1$ MPa. The structure is split into 8 bodies and each of them consists of 2 finite elements (see Fig.1). All bodies are considered as variables ($N2 = N2C = 8$) and their thicknesses can be taken from the sample $B2(I, IZ) = 0.6, 0.8, 1., 1.2, 1.4, 1.6$ mm identical for all $I = 1$ to 8 ($IZM = 6$). The SZ function was chosen in form $SZ(I, IZ) = 1., 1.1, 1.2, 0.8, 0.4, 0.1$ (also identical for all $I = 1, 8$). $DELT = 1.2$, $BK = 0.5$.

The following tables give the solution $IX(I)$ in individual iterations IT (the first solution is the chosen starting one). The solution denoted as F.D. is the feasible solution. The last one from the sequence of $IX(I)$ with F.D. is the required extremum.

Example 1 $ID = 1$

5	6	5	6	5	6	5	6	FD
4	5	4	5	4	5	4	5	FD
3	4	3	4	5	6	5	6	FD
2	3	2	3	4	5	4	5	FD
1	2	1	2	3	4	3	4	FD
1	1	1	1	2	3	2	3	FD
1	1	1	1	2	1	2	1	
1	2	2	2	2	3	2	3	
1	2	2	2	2	3	2	3	
1	1	1	1	2	1	1	2	

Example 2 $ID = 1$

1	2	1	2	1	2	1	2	
2	3	2	3	2	3	2	3	FD
1	2	1	2	3	2	3	2	
1	2	1	3	4	3	4	3	
1	2	1	2	3	4	3	4	
1	1	1	1	2	3	2	3	FD

The solution of Ex.1 is therefore $IX(I) = 1\ 1\ 1\ 1\ 2\ 3\ 2\ 3$ i. e. in the body $I = 1$ is its variant $IZ = 1$, in body $I = 2$ its variant 1 etc. The solution of Example 2 is the same.

6. Conclusion

The described treatment is similar because of (26) for $G(I1) = 1$ and SZ for $ID = 1$ to the gradient treatment. For $\frac{\partial G(I1)}{\partial b_1} = 1$ and $SZ = 1$ it is close to the treatment by the *full stress method*. The generalized factor in (26) is the SZ function and here a series of open problems arises. There evidently can be chosen a series of tentative hypotheses how to choose SZ - with the aim to accelerate the convergence or to increase the probability to attain the global extremum. One possibility is to choose several patterns of SZ function with greater or lesser slope of the increase and with the greater or lesser scope (ID). Then the program itself can chose individual SZ at an individual constraints I1: for example by $G(I1) >> 0$ and $\frac{\partial G(I1)}{\partial b_1} < 0$ to use SZ with a larger gradient and with a widez scope. At $G(I1) > 0$ then SZ with lesser scope etc.

References

1. J. Farkas, Optimum design of metal structures. Akademia Kiàdo, Budapest, 1984

2. E. J. Hung and S. J. Arora, Applied optimum design. J. Wiley, London, 1979

3. R. T. Haftka and A. M. Kamat, Elements of structural optimization. Kluver, Dorthecht, 1984

4. H. Gallagher and O. C. Zienkiewicz, Optimum structural design. J. Wiley London, 1973

5. J. Mottl, Description of a program for nonlinear programming. Computer J. 22, No.3 1979

6. J. Mottl, Truss system optimization using the voting method. Computer & Structures vol. 45, No.1, p.127-149, 1992

7. G. V. Reklaitis, A. Ravindran, K. M. Ragsdal, Engineering optimization. J. Wiley, N.Y. 1983

8. J. Ratschek, J. Rakve, New computer methods for global optimization. Ellis Haarwood 1988

9. S. S. Rao, The finite element method in engineering. Springer Verlag, Berlin 1982

10. C. N. Vanderplaats, Numerical optimization technique for engineering design. Mc Graw-Hill, N.Y. 1984

Neuro-Optimizer, Its Application to Discrete Structural Optimization

KISHI Mitsuo, KODERA Toshiyuki, IWAO Yoshiyuki, and HOSODA Ryusuke

Department of Marine System Engineering, College of Engineering, University of Osaka Prefecture, Sakai, Osaka 593, JAPAN

Abstract. Some discrete optimization problems can be programmed and solved on artificial neural networks. The NEURO-OPTIMIZER attains good (not necessarily optimal) solutions for general nonlinear discrete optimization problems through neurons state transitions. The simulated annealing method is introduced to escape from local minima. In this paper the number representation of the discrete variable by using neurons is investigated, and a mapping technique is proposed for irregularly discrete variables.

Keywords.   discrete optimization, neural network, structural optimization, number representation

1 Introduction

In the middle 1980's Hopfield [1] showed that some combinatorial/discrete optimization problems can be programmed and solved on artificial neural networks minimizing the quadratic energy function. The basic concept of the Hopfield neural network is a combination of the input-output neuron model and the steepest descent method.

Based on the concept of the Hopfield neural network we have already proposed an optimization method named NEURO-OPTIMIZER [2] which is expected to be able to attain good solutions for general nonlinear discrete optimization problems. The NEURO-OPTIMIZER needs some algorithm to escape from local minima of the energy function. The simulated annealing method [3] is introduced here.

The discrete variable is represented numerically by neurons in the NEURO-OPTIMIZER. The neuron state takes binary values of one or zero. In this paper the number representation of the discrete variable is investigated by a stability analysis, and a mapping technique is proposed as a redundant representation for irregularly discrete variables. Numerical examples for structural optimization are provided to illustrate the applicability of the NEURO-OPTIMIZER with the mapping technique.

2 NEURO-OPTIMIZER

2.1 Neuron Model

The neural network consists of mutually interconnected neurons. In the NEURO-OPTIMIZER based on the concept of the Hopfield model, the system behaviour is formulated with state equations as follows:

$$V_i(t) = \Phi[U_i(t)] \tag{1}$$
$$dU_i(t)/dt = -\partial E/\partial V_i \tag{2}$$

where $V_i(t)$ is the state/output of neuron i at the instant t, U_i is the input from other neurons, and E is the energy function defined for the neural network. Input-output function Φ is a monotonic increasing function which takes values between zero and one, for example, in the following form:

$$\Phi(u) = 1/(1 + e^{-u}). \tag{3}$$

Energy function E must be differentiable with respect to the neuron state V_i. Then the neurons change their state decreasing the energy E, i.e. $dE/dt \leq 0$.

2.2 Discrete Optimization

Consider a discrete optimization problem such as:

Find	X	
Such that	$f(X) \rightarrow$ minimize	(4)
Subject to	$g(X) \leq 0$	(5)
	$h(X) = 0$	(6)
	$X \in \chi$	(7)

where X is the discrete variable, f is the objective function, g and h are the constraint functions, and solution space χ denotes the finite set of all possible solutions.

The Lagrangean function L is defined as:

$$L(X, \lambda, \mu) = f(X) + \lambda g(X) + \mu h(X). \tag{8}$$

$\lambda (\geq 0)$ and μ are the Lagrange multipliers. The optimal solution X^* can be obtained minimizing L with respect to X $(\in \chi)$. The Lagrange multipliers λ and μ are improved in the following way [4]:

$$\Delta \lambda \propto g(X) \tag{9}$$
$$\Delta \mu \propto h(X) \tag{10}$$

where Δ denotes the increment. Negative λ is treated as zero.

Suppose that the discrete variable $X(\in \chi)$ can be represented by neurons state V, and let the energy function E of the neural network be the Lagrangean function L. Then the neurons change their state decreasing the Lagrangean. This means that the above mentioned discrete optimization problem can be solved on the neural network. The equality constraint $\mathbf{V}^\tau(\mathbf{1}-\mathbf{V})=0$, where $\mathbf{1}$ denotes the vector whose elements are unity, is added for the binarization of V so that each neuron i has a state: $V_i=0$ or $V_i=1$.

2.3 Fluctuating Neuron Model

The discrete optimization problem involves local optima. The energy function of the neural network has a very large number of local minima because of the binarization constraint on the neuron state. Some fluctuating neuron models have been proposed in order to escape from local minima.

One approach to fluctuating neurons is to use a probabilistic state transition mechanism. The Boltzmann machine [3] is a probabilistic model, where the input-output function Φ gives the probability of the neuron state V_i being one, i.e.:

$$\Phi(u) = 1/(1 + e^{-u/T}) \tag{11}$$

$$\Phi(U_i) = \text{Prob}[V_i=1]. \tag{12}$$

The parameter T is the so-called temperature which is controlled in an annealing schedule.

The Langevin equation model [2] is another probabilistic model, in which the random noise is added to the input U_i of the form:

$$dU_i(t)/dt = -\partial E/\partial V_i + u_i. \tag{13}$$

The noise u_i is the zero-mean normally distributed random variable, and the variance $\sigma_{u_i}^2$ is given by

$$\sigma_{u_i}^2 = \alpha T^2 \tag{14}$$

where α is a positive constant, and T is the temperature.

The other approach to fluctuating neurons is to use a chaotic dynamics. State equations of neural networks have an affinity for the chaos. That approach is attractive; however, at present much investigation should be done in order to control chaotic system parameters.

In the following, we introduce the Langevin equation model because of its simplicity. The temperature is slowly lowered in the annealing schedule, for example:

$$T(q) = T_0/\ln(1+q) \tag{15}$$

where $T(q)$ denotes the temperature during the q-th stage, and T_0 is some positive constant.

3 Number Representation of Discrete Variable

3.1 Number Representations

In the NEURO-OPTIMIZER each discrete variable is represented by neurons which take binary values of one or zero. There are various ways of representing discrete variables numerically, for example:

(A) Distributed representation - For the variable X which takes regularly discrete values, e.g. d, d+c,..., d+(n-1)c, the distributed representation is introduced of the form:

$$X = d + c\sum_i V_i \qquad (16)$$

where the sum is over n-1 values, and each neuron is independent.

(B) Local representation - For the variable X which takes irregularly discrete values, e.g. $d_1,...,d_n$ ($d_i < d_{i+1}$), the local representation is introduced of the form:

$$X = \sum_i d_i V_i, \qquad \text{subject to } \sum_i V_i = 1 \qquad (17)$$

where the sum is over n values, and each neuron is exclusive.

3.2 Stability Analysis

From Eqs.(1) and (2) we have

$$dV_i/dt = -V_i(1-V_i)\partial L/\partial V_i \triangleq p_i(V). \qquad (18)$$

Let $\underline{V}$ be a neurons state vector whose elements take binary values of one or zero. By expanding Eq.(18) in a Taylor series and retaining only the linear terms we obtain

$$dV_i/dt \approx p_i(\underline{V}) + (\partial p_i/\partial V|_{\underline{v}})^\tau \Delta V = (2\underline{V}_i-1)(\partial L/\partial V_i|_{\underline{v}})\Delta V_i$$
$$\triangleq q_i\Delta V_i. \qquad (19)$$

Thus, it is natural to arrive at the following conclusions for the neural network system subjected very small fluctuations: i) If $q_i<0$ for any i, the system is stable. ii) If $q_i>0$ for some i, the system is unstable.

Simple example of the stability analysis is now given. Consider an optimization problem such as:

$$\begin{array}{lll} \text{Find} & X & \\ \text{Such that} & f(X) \rightarrow \text{minimize} & (20) \\ \text{Subject to} & X \in \chi=\{r_1,...,r_k^*,...,r_n\} & (21) \end{array}$$

where X is a scalar variable, and f is a convex function whose minimal point

26

is $X = r_k^*$. Let $r_1,..., r_n$ be an ordered series with equal difference c. Thus, the distributed representation can be used here. The Lagrangean function is defined as:

$$L(V, \mu_1) = f(r_1 + c\sum_{i=1}^{n-1} V_i) + \mu_1 \sum_{i=1}^{n-1} h_i(V_i). \qquad (22)$$

The constraint functions $h_i(V_i) = V_i(1-V_i)$ $(i=1,...,n-1)$ are added for the binarization of neuron state V_i, and μ_1 is the Lagrange multiplier. We want to let the optimal solution $X = r_k^*$ be stable and the other binary solutions be unstable. As a result of the stability analysis, we have the following feasible bounds on the Lagrange multiplier μ_1:

$$0 < \mu_1 < \min_{X \in \chi,\ X \neq r_k^*} |cf'(X)| \qquad (23)$$

where f' indicates the derivative of f.

On the other hand the local representation is used for the discrete variable X. Then the Lagrangean function is defined as:

$$L(V, \mu_1, \mu_2) = f(\sum_{i=1}^{n} r_i V_i) + \mu_1 \sum_{i=1}^{n} h_i(V_i) + \mu_2 (\sum_{i=1}^{n} V_i - 1)^2 \qquad (24)$$

where μ_2 is the Lagrange multiplier for the constraint function in Eq.(17). As a result of the stability analysis, we found that it is impossible to let only the optimal solution be stable.

3.3 Mapping Technique

The industrial standard often involves irregularly discrete values. The distributed number representation is desirable even for irregularly discrete design variables in the NEURO-OPTIMIZER. Here we propose a mapping technique in order to transform the sequence of irregularly discrete values to a sequence of regularly discrete values.

Suppose that the variable X takes irregularly discrete values, $d_1,...,d_n$. We introduce a mapping F by which integer i-1 $(i=1,...,n)$ has a one-to-one correspondence to discrete value d_i (see Fig. 1). Hence, the optimization problem with respect to X is transformed to the optimization problem with respect to the integer variable Y by the mapping $F: Y \rightarrow X$. The spline

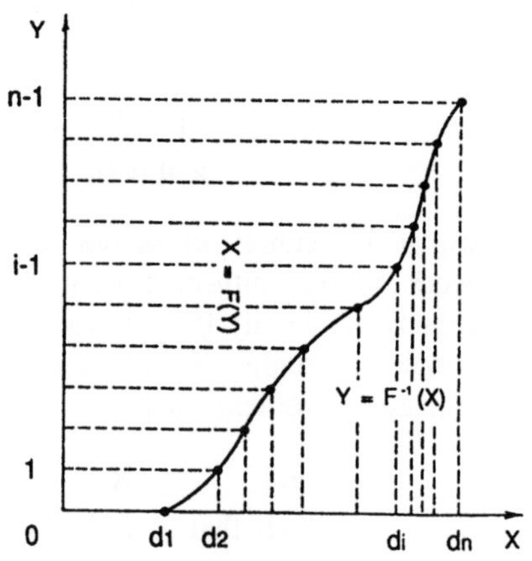

Fig. 1. Mapping technique

function or the Lagrange interpolation can be used in order to represent F. Then, the integer variable Y is represented by Eq.(16) with binary neurons V, and the NEURO-OPTIMIZER can solve the optimization problem.

The mapping technique is applicable when the discrete variable takes linear ordered values. Conversely, for the discrete variable which takes unordered values, the technique is not proper.

4 Application to Structural Optimization

Numerical examples for structural optimization are presented to demonstrate the applicability of the NEURO-OPTIMIZER. The structural members are uniform and homogeneous. The failure criteria are deterministic. The configuration of the structure, the materials to be used, and the loading conditions are assumed to be given. The size of each member is singled out of the candidates.

4.1 Optimal Design of Ship Structure

The optimal design problem is solved for the ship structure as shown in Fig. 2. The design problem is to determine the optimal size of the longitudinal members which minimize the cross-sectional area of the midship under the constraint of structural section modulus. The rule requirement by the ship classification society is adopted. The member candidates satisfy the rule requirement of the local strength.

The number of design variables is twenty-three considering the members grouping, and each design variable has five candidates. The mapping technique is used to represent the discrete design variables. The state transition of the neural network is simulated by using the Euler method. Neurons state transitions occur synchronously. The temperature and the Lagrange multipliers are renewed at every iteration. The objective function and the constraint functions are scaled to keep the input increment $dU_i/dt\Delta t$ at appropriate level($=O(10^{-1})$). The numerical data are given as $\alpha=1/\pi$ in Eq.(14), $T_0=20.0$ in Eq.(15), and the simulation time step $\Delta t=0.05$. Every Lagrange multiplier takes zero-initial-value. The initial condition of the neurons is generated randomly.

For the design problem one hundred simulations are carried out. On the other hand twenty thousand feasible solutions are found by the Monte Carlo method. The NEURO-OPTIMIZER finds solutions which are better than the best solution by the Monte Carlo method with 77% probability of success.

4.2 Optimal Design of Truss Structure

The optimal design problem is solved for the truss structure as shown in Fig. 3 [6]. The design problem is to determine the optimal cross-sectional size of the members minimizing the structural weight/mass under the yielding and buckling constraints [5].

28

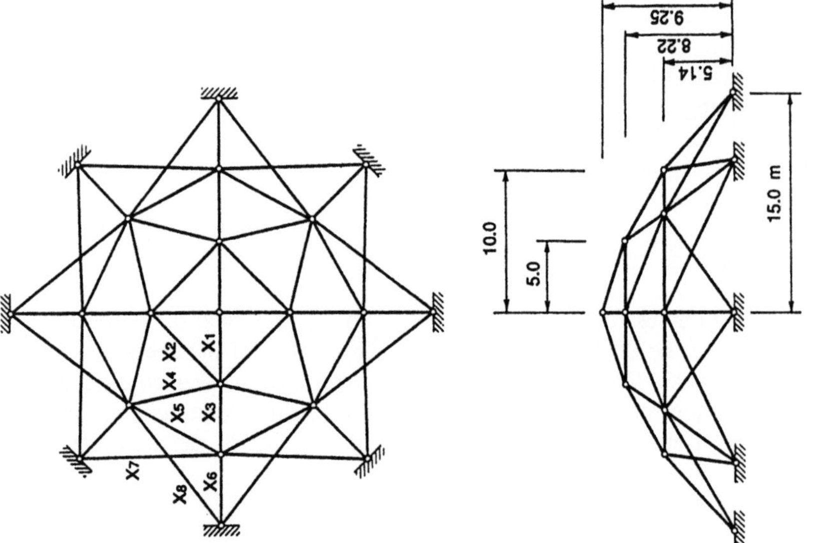

Fig. 3. Truss structure

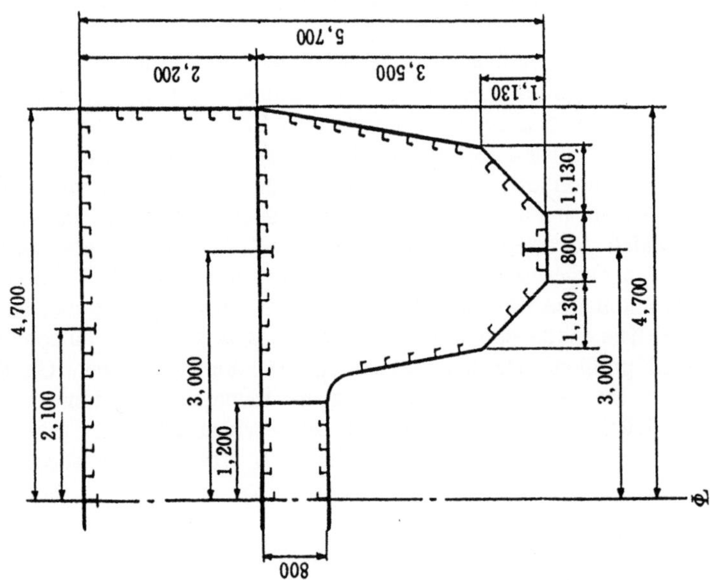

Fig. 2. Midship section

The material data and the loading conditions are given following the reference [6]. The number of design variables is eight considering the symmetric structure, and each design variable has ten candidates.

For the optimal design problem fifty simulations are carried out. The optimal solution [6] is found with 48% probability of success. The average number of state transitions of the neural network until convergence is 124.7. Although the NEURO-OPTIMIZER does not guarantee to attain the optimal solution, the quality of the obtained solutions is satisfactory.

5 Conclusions

This paper is concerned with the discrete optimization method, NEURO-OPTIMIZER. The number representation of the discrete design variable by neurons is investigated through a stability analysis, and the mapping technique is proposed as a distributed representation. The NEURO-OPTIMIZER is applied to discrete structural optimization problems, and good solutions are found with a high probability of success.

This paper is silent on scaling mechanisms of the Lagrangean function; however, the scaling is related to the following two quantities: i) the quality of the solution obtained by the NEURO-OPTIMIZER, and ii) the computational time until convergence. Actually the scaling is problem-dependent, the general mechanism should be proposed.

A part of this work is financially supported by a Grant-in-Aid for the Scientific Research, the Ministry of Education, Science and Culture of Japan. All the computations were processed by using the NEC PC-9800 system.

References

1. Hopfield, J.J., Electronic Network for Collective Decision Based on Large Number of Connections between Signals, United States Patent, 1987, No.4,660,166.
2. Kishi, M., Suzuki, T., and Hosoda, R., Structural Design by Neuro-Optimizer, in Practical Design of Ships and Mobile Units, eds. J.B. Caldwell and G. Ward, Vol.2, Elsevier Applied Science, 1992, 940-952.
3. Aarts, E., and Korst, J., Simulated Annealing and Boltzmann Machines: A Stochastic Approach to Combinatorial Optimization and Neural Computing, John Wiley & Sons, 1989.
4. Platt, J.C., and Barr, A.H., Constrained Differential Optimization, in Neural Information Processing Systems, ed. D.Z. Anderson, American Inst. of Physics, 1988, 612-21.
5. Pedersen, P., On the Optimal Layout of Multi-Purpose Trusses, Comput. Struct., Vol.2, 1972, 695-712.
6. Pedersen, P., Optimal Joint Positions for Space Trusses, J. Struct. Div., ASCE, Vol.99, St10, 1973, 2459-2477.

Genetic Algorithms in Topological Design
of Grillage Structures

P. Hajela and E. Lee
Mechanical Engineering, Aeronautical Engineering & Mechanics
Rensselaer Polytechnic Institute, Troy, New York, 12180 USA

Abstract. The present paper describes the use of genetic algorithms (GA's) in determining the optimal layout and sizing of grillage structures for stress and displacement constraints. The design space for this problem is highly nonconvex, and not readily amenable to traditional methods of nonlinear programming. The approach develops an optimal topology from a set of predefined structural elements so as to satisfy kinematic stability requirements in addition to the usual constraints of structural strength and stiffness. A two-level GA based search is used, wherein the kinematic stability constraints are imposed at one level, followed by the treatment of stress and displacement constraints at a second level of optimization. Since GA's search for an optimal design from a discrete set of alternatives in the design space, their adaptation in the topological design problem is natural, and is governed only by issues related to computational efficiency. Strategies designed to alleviate the computational requirements of a GA based search are discussed in the paper.

Keywords. Genetic algorithms, grillage topological optimization, discrete design variables

1. Introduction

In pursuing optimal design of flexural systems, one has to contend with problems associated with multimodal design spaces, and a force distribution that is highly sensitive to the cross-sectional dimensions [1-3]. These problems are further compounded when the topology of the structure is allowed to change during the design process, as members can be removed or added to the structural system. Kirsch [4] used the force method formulation in the topological design of grillage structures, as it facilitated a study of the influence of compatibility conditions on the optimal design. In the equilibrium linear programming approach (ELP) used in that work, the compatibility conditions were ignored, making it possible to state the stress constrained problem as one of linear programming, with the redundant forces in the members as design variables. While this approach works well for the class of stress constrained problems, its extension to more general response constraints is not readily apparent. Rozvany [5] proposed the use of the optimality criteria approach in this problem. For multiple response constraints, the solution splits up into distinct regions, each governed by a different optimality criterion. Partial resolution of this problem was obtained by use of layout theory, as shown in Reference 6.

While the nonlinear programming approach can be used to address the problem of multiple response constraints, its effectiveness is limited to smaller problems, as an increase in the number of design variables (increasing the number of admissible structural elements) is detrimental to the search efficiency. The present study was motivated by the need to develop a more general approach for structural topology development in the context of static and dynamic structural response constraints. These include structural topology determination and sizing for strength, stiffness, dynamic response, and energy dissipation capacity.

A genetic algorithm based global search strategy for generating near-optimal structural topologies is proposed, and may be considered a derivative of the ground structure approach used in the topological design of truss structures [7]. The application of this approach is particularly potent, in that structural members can be both added and removed during the search process. The method is applicable to general design constraints, and has been shown to offer an increased probability of locating the global optimum in the presence of nonconvexities in the design space [8]. Subsequent sections of the paper discuss an implementation of this strategy in the topological design of grillage structures. The principal concern resides in selecting strategies that reduce the overall computational requirements.

2. Problem Statement

The approach used in the present work is a derivative of the ground structure approach, where the ground structure specifies all possible structural elements and connectivities. The role of the optimizer is to retain only those elements that would yield a minimal weight design that also meets the design requirements. A general mathematical statement of this problem is as follows:

$$\text{Minimize } F(Y) = \sum_{j=1}^{J} i_j \rho \, l_j Y_j$$

Subject to:

$$g_k(Y) \leq 0 \qquad k = 1, K$$

$$h_p(Y) = 0 \qquad p = 1, P$$

$$Y_j^L \leq Y_j \leq Y_j^U$$

Here, J is the total number of structural elements that can be used in establishing the topology; i_j are components of a J-vector I, and a value of 1 or 0 indicates the presence or absence, respectively, of the j-th structural member; Y_j is the cross-sectional area of the j-th element, and superscripts 'L' and 'U' indicate lower and upper bounds, respectively; ρ is the material density and l_j is the length of the j-th member; h_p and g_k are equality and inequality constraints on the response quantities. In the problem under consideration, only inequality constraints on stress and displacements are considered, where these response quantities are obtained in a displacement based finite element analysis. Additional requirements on structural geometry may also be stipulated in the constraint set. Such requirements may include, for example, that all support points be utilized in the load bearing process. Addition or removal of elements from the ground structure can result in kinematically unstable structures. In a displacement based analysis formulation, the requirement of a positive definite stiffness matrix would be violated as a result of this instability.

A two-stage optimization process was adopted, involving topological design for kinematic stability requirements at the first level (S1). At the second stage (S2), both member resizing and addition/removal of members was considered. As shown in subsequent sections of the paper, genetic algorithms are particularly well suited for this task.

3. Genetic Search

Genetic algorithms, first proposed by John Holland in 1975 [9], have been adapted for a number of applications involving optimal design. The method can be best described as belonging to a general category of stochastic search techniques, and has its philosophical basis in Darwin's theory of survival of the fittest. A set of design alternatives which represent a generation in the natural analoque are allowed to reproduce and cross among themselves, with bias allocated to the most fit members of the population. Combinations of the most desirable characteristics of the mating pairs of the population result in progenies that are more fit than either of the parents. If the measure which indicates the fitness of the population is also the desired goal of the design process, successive generations will result in better objective function values. While a detailed description of the method is beyond the scope of the present paper, the basic operations are included here for completeness.

The process is initiated with the random generation of an initial population of candidate designs; very often, this pool is enhanced by the introduction of a number of known good designs. A fitness function based on the objective function and constraint values of each design is evaluated (in present work, constraint violations were appended to the function to be minimized in a penalty function format), and on the basis of this fitness, each design was assigned a probability of being selected as a mating parent. Genetic algorithms apply selection pressure through this process, allowing designs that are more fit in one generation to increase their presence in subsequent generations. In a fixed population simulation of the search, designs with lower fitness are eliminated from subsequent generations. An elitist selection strategy is often applied, wherein a fixed number of best designs is always retained from one generation to another. Once a mating pool is selected, the crossover operation is performed. Crossover is the primary mechanism to introduce new designs into the population. The traditional two point crossover operation involves selecting two crossover sites on each string at random, and interchanging the binary substrings bracketed by the crossover sites between the mating strings. The population is finally subjected to the mutation operation, which constitutes random selection of a bit location on a string, and changing the 0 to a 1 or vice versa; this operation is performed with a low probability (less than 2% of the designs are typically affected). These operations comprise one generation of genetic evolution. The genetic search is typically terminated when the best design does not improve over a specified number of generations or when a chosen maximum number of function evaluations have been performed.

The method does not require the use of gradients of objective and constraint functions, and is therefore potent in problems where the design space may be nonconvex or discontinuous. Genetic search uses a chromosome-like binary string to represent an actual design. The physical design variables are first mapped into binary strings, genetic transformation operations are performed on these strings, and the physical variables recovered in an inverse transformation for function evaluation. It is obvious that a binary string of finite length is only able to represent a finite number of design alternatives. While this characteristics may require long strings to adequately represent continuous variables, it is ideally tailored for the representation of discrete or integer variables. Finally, in genetic search, information from all members of a current generation is used to update the population. Since these members are distributed all over the design space (at least initially), there is a reduced possibility of convergence to a relative optimum.

In the present application, the representation of design variables by binary strings is relatively straightforward. For the stage S1, a binary string of length J given by the total number of admissible elements was used. A position on the string corresponds to a particular structural element, and a value of 0 or 1 indicates the absence or presence of the element in the topology. When the number of elements in the grillage increases, this string can become quite large. In the S2 problem, where member sizing is included, the length of the binary string is a function of the number of elements, and the precision with which each element is to be represented. Since an M-digit binary string is capable of generating 2^M distinct combinations, the precision A_c with which a variable Y_j can be represented is given by $A_c = (Y_j^U - Y_j^L)/(2^M - 1)$. Note that a higher precision requires that either the range of design variable variation be reduced, or the string length M be increased. For large sized problems, where Y_j are to be considered continuous variables, the string lengths can be quite large. Larger string length imply a larger number of design alternatives for the GA to evaluate, and a larger population size is required to perform adequate sampling at each generation of evolution.

In the present work, the genetic search code EVOLVE [10] was used for optimization, which includes a number of advanced strategies in addition to the basic operations described above. Among these strategies is the directed crossover (replaces random crossover by using generational record of how crossover at certain sites performed) and a multistage search. The latter successively increases the precision of design variable representation without having to increase the population size as would be otherwise required. Another useful feature in this code is an implementation of a sharing function strategy. Use of this strategy allows the location of multiple relative optima in the design space in a single trial of GA search. Since the sharing approach is based on preventing all members of the population from converging to a single optimum point, it was used in a rather unique way in the present application to maintain diversity in smaller sized populations for a larger number of generations, preventing premature convergence, and allowing more design alternatives to be evaluated during the search process. These advanced strategies are more beneficial as the problem dimensionality is increased.

4. Description of Test Problems

The GA based search was implemented in two test problems - a three beam grillage and a ten beam grillage structure as shown in Figures 1 and 2. These problems were studied in the context of stress constraints in Reference 4, where an Equilibrium Linear Programming approach (ELP) was used to derive the optimal topologies. In the present numerical work, the designs obtained were compared to those in [4] and were extended to include displacement constraints. For each of the two problems presented here, arbitrary units were assumed to simplify the discussion.

4.1 Three-Beam Grillage

For this structure, two concentrated loads of magnitude P=10.0 were applied as shown. Note that this choice of applied loads allows elimination of either the longitudinal or the transverse beam elements. A rectangular cross-section was prescribed for the beam elements with a constant width of 12.0; the depth of the longitudinal and transverse beams, denoted by X_1 and X_2, respectively, were the design variables for the problem. Geometry of the grillage was defined by choosing l_y=1.4 and l_x=1.0. Excluding the torsional deformations of the beam elements, and considering only the stress constraints with allowable stress levels of σ_{al}=1.0, the design space for the problem is as shown in Figure 3. Even in this simple problem, the design space is nonconvex due to the presence of multiple load paths in the structure. Three distinct topologies could be candidates for the optimal design, and are identified as follows:

a) A determinate structure in the l_x direction with X_2=0 (denoted as Z_1^*)

b) A determinate structure in the l_y direction with X_1=0 (denoted as Z_2^*)

c) An indeterminate structure with nonzero X_1 and X_2 (denoted as Z_3^*)

While this problem offers only a small number of possible topologies, for even a moderate increase in the problem size, the number of relative optima would be large and not amenable to exhaustive enumeration.

4.2 Ten-Beam Grillage

For this structure, a total of 21 concentrated loads (each P=10.0) are to be carried at the node points indicated in Figure 2. A total of 3 longitudinal and 7 transverse beam elements can be used for this purpose, and a number of different geometric variations were considered. These variations were introduced by a parameter $\alpha=l_y/l_x$, where the lengths l_y and l_x were also constrained as $l_y l_x$=1. As in the previous example, torsional deformations of the beam elements were excluded. A rectangular cross-section of the beam elements was again assumed; a fixed depth was specified, and the widths of the beam cross-section were considered as the design variables. For this choice of design variables, the bending moments in the beam are proportional to the cross-sectional area, and it is easier to vizualize the effects of the bending moment distribution on the topology of the grillage structure. For this test problem, the values of stress and displacement allowables were specified as σ_{al}=1.0 and w_{al}=0.3, respectively, were used; a Young's modulus value of E=1.0 was used in computing the displacements. Note that this topology is symmetric with respect to the x-y axes system, and a quarter model shown in Figure 2a was employed for all analysis.

5. Discussion of Results

For the three-beam grillage structure, the GA based solution to S1 problem was trivial, and identified each of the three possible stable topologies. Here, a 2 digit binary string was sufficient to represent the design variables; the first digit pertained to the longitudinal member and the second digit corresponded to the transverse member. A 0 or 1 implied the absence and presence, respectively, of the structural member corresponding to that location. A small population size of 15 was sufficient for this stage of the solution. For the S2 problem, where member sizing was considered in addition to the topology identification problem, the sharing function approach was used to locate each of the three relative optima in a single GA based search. In this search, two different population sizes of N=120 and N=210 were used; probabilities of crossover and mutation, p_c=0.8 and p_m=0.02, were selected. The results obtained by Kirsch [4] are compared with the two GA trials in Table 1. Note that the results of Reference 4 are based on a continuous variation of the design variables X1 and X2, while the GA results are based on a discrete representation of these variables with a precision of 0.02. To achieve this precision, a 7 digit binary string was used to represent each design variable, where upper and lower bounds of 2.54 and 0.0 were specified for the latter. When the sharing function approach was invoked to locate multiple relative optima, each of the three locally optimal designs were identified. Without the use of the sharing function, the globally optimal design was found.

A number of different experiments, including problems with stress constraints only as in [4], and a combination of stress and displacement constraints, were conducted with the ten-beam grillage structure. Symmetry of load and support conditions allows a reduction in the number of design variables to 6, as is obvious from the quarter model of Figure 2a. The probabilities of crossover and mutation were retained from the previous example for the set of numerical experiments. A population size of N=30 was used in the S1 problem. In this stage, variables X_i, i=1,6, were assigned nominal values of 1.0. A 6 digit binary string was used to denote the presence or absence of the structural members. A total of 19 stable topologies are possible , and each was identified in the first level of search. For each of these 19 stable topologies, 8 different combinations of design variables were randomly generated to create the initial population of size N=152 for the S2 problem. Note that this choice attempts to preclude any bias in the starting population. For the S2 problem, each design variable was represented with a precision of 0.1, which for design variable lower and upper bounds of 0.0 and 6.3, respectively, required each variable to be represented by a 6 digit binary string. Hence, a 36 digit binary string was required to represent the grillage design. Note that this string length implies that the GA is searching for an optimal design from 2^{36} possible designs, not all of which are feasible.

For the stress constrained problem, several values of α were considered, and a summary of these results is shown in Table 2. The table shows the optimal weights for the assumed rectangular cross section of the beams. In Reference 4, the objective function was obtained as $F'=l^T M_m$, where l is the length of any segment of the grillage beam, and M_m is the maximum bending moment for that segment. For each of the optimal designs obtained in the present study, and for the assumed cross section, a corresponding value of F' is presented in Table 2 for comparison with results of [4]. Note that for statically determinate structures, where cross section dimensions do not figure in computing F', the results are identical. Note that good agreement exists with results presented in [4]. In some cases, as for α=1.2, the topologies are slightly different. Member X_1, however, is small, and hence the slight difference in the objective function value. It is important to bear in mind that the optimal topologies obtained in this work can also be different from those of Reference 4 due to the fact that discrete variations in member dimensions were admitted in the present work.

In the ELP approach used in [4], it is not possible to include displacement constraints in the design problem. This presents no additional challenge in the present approach, and a constraint on vertical displacement under each of the applied loads was included in the problem statement. The design variables were each encoded by a 7 digit binary string for this problem, as an expanded range of variation of design variables had to be considered. The lower and upper bounds on the variables were 0.0 and 12.7, respectively. A summary of results for α=1.0, 1.5, 2.0, and 2.5 is presented in Table 3. For α=1.0, the optimal topology is the same as that obtained for stress constraints only, with the weight increasing by a small amount to a value of 364.0. When a continuous variation in design variables was admitted for this problem in a NLP based search, the weight was reduced to 354.67. For α=1.5, a statically indeterminate topology was obtained. While it was not possible to confirm if this is indeed the optimal design, several other topologies generated in the GA search were subjected to a NLP based design optimization. Although no lower weight topology was identified, it is still possible that the optimal design was not found due to the nonconvexities in the design space for the member sizing problem. For values of α=2.0 and 2.5, the topologies identified as optimal were statically indeterminate. As shown in Table 3, for α=2.0, members X_4 and X_5 were added to the optimal grillages obtained for stress constraints only (Table 2). Similarly, the optimal topology for α=2.5, for stress constraints only, was a determinate structure. If that determinate structure is simply stiffened to account for displacement constraints, an optimal objective function value of 819.66 is obtained. This compares to an optimal weight of 750.1 obtained by going to an indeterminate structure shown in Table 3.

It is interesting to note that the largest increment in the structural weight obtained for stress constraints due to the inclusion of displacement constraints is for α=2.0. This ratio results in maximum lengths for both the longitudinal and transverse members, and since deflection at the mid point of these members is proportional to the cube of their lengths, greater stiffening is required to meet the requirements imposed by the displacement constraint.

6. Closing Remarks

The paper describes an application of genetic algorithms to the topological design of grillage structures. This problem has a highly nonconvex design space, which effectively precludes the use of the more traditional mathematical programming methods for the most general problem. Genetic algorithms which work effectively with discrete variables, and which do not require computation of the response gradients, are shown in the present work to be effective in locating near-optimal designs. Numerical results obtained for two grillage design problems are presented in support of the proposed approach. An extension of the approach to problems with larger number of design variables is presently being pursued. This effort uses specialized strategies of directed crossover and multistage search to circumvent an inordinate increase in computational effort.

7. Acknowledgements

Support received under ARO contract DAAH-04-93-G-0003 is gratefully acknowledged.

8. References

1. D. Kavlie and J. Moe, "Application of Nonlinear Programming to Optimal Grillage Design", International Journal of Numerical Methods in Engineering, 1, 351-378, 1969.

2. F. Moses and J. Onoda, "Minimum Weight Design of Structures With Application to Elastic Grillages", International Journal of Numerical Methods in Engineering, 1, 311-331, 1969.

3. U. Kirsch and A. Taye, "On Optimal Topology of Grillage Structures", Engineering with Computers, 1, 229-243, 1986.

4. U. Kirsch, "Optimal Topologies of Flexural Systems", Engineering Optimization, Vol. 11, pp. 141-149, 1987.

5. G.I.N. Rozvany, "Optimality Criteria and Layout Theory in Structural Design: Recent Developments and Applications", in Structural Optimization, eds G.I.N. Rozvany and B.L. Karihaloo, pp. 265-272, 1988.

6. G.I.N. Rozvany, "Optimality Criteria for Grids, Shells and Arches", in Optimization of Distributed Parameter Problems, eds. E.J. Haug and J. Cea, pp. 112-151, 1981.

7. P. Hajela, E. Lee and C.-Y. Lin, "Genetic Algorithms in Structural Topology Optimization", presented at NATO Advanced Research Workshop, Sesimbra, Portugal, June 22-28, 1992, to be published in Topology Design of Structures, eds. M.P. Bendsoe and C.A. Mota-Soares, Kluwer Academic, 1992.

8. P. Hajela, "Genetic Search - An Approach to the Nonconvex Optimization Problem", AIAA Journal, Vol. 26, No. 7, pp1205-1210, July 1990.

9. J.H.Holland, Adaptation in Natural and Artificial Systems, The University of Michigan Press, Ann Arbor,1975.

10. C.-Y. Lin and P. Hajela, "EVOLVE: A Genetic Search Based Optimization Code With Multiple Strategies, proceedings of OPTI'93, Computer-Aided Optimum Design of Structures, 7-9 July, Zaragoza, Spain, 1993, pp. 639-654, Elsevier Applied Science, London, eds. S. Hernandez and C.A. Brebbia.

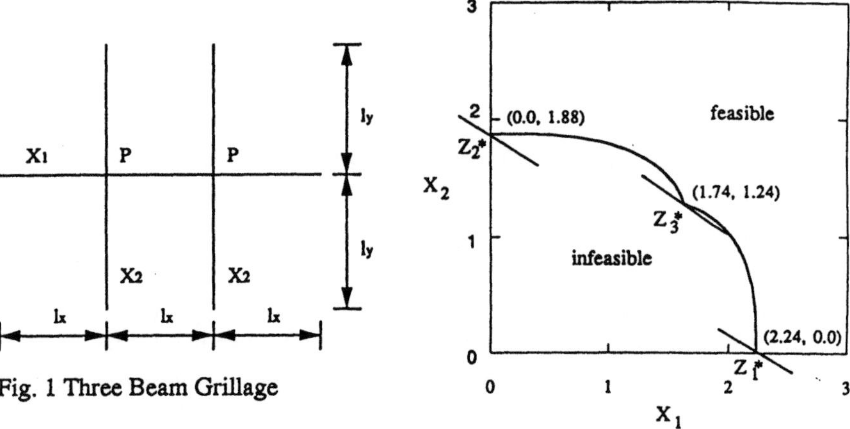

Fig. 1 Three Beam Grillage

Fig. 3 Design Space for Three Beam Grillage

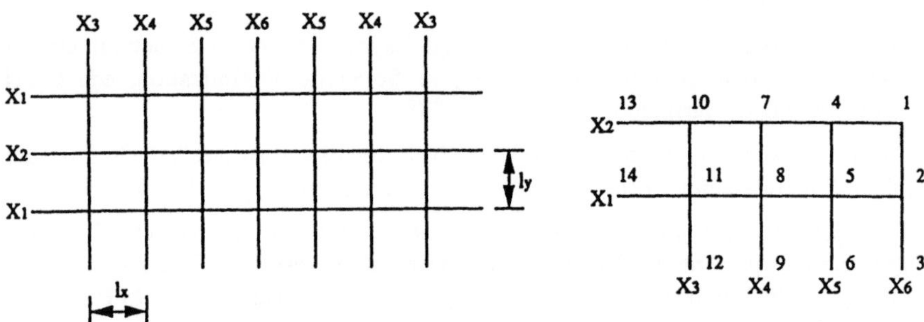

Fig. 2a Ten Beam Grillage

Fig. 2b Quarter Model of Ten Beam Grillage

Table 1. Results for Three Beam Grillage (stress constraints only)

Design Identifier	Results of Reference 4 Δ = continuous			Results of Present study $\Delta = 0.02$		
	X_1	X_2	Objective fn.	X_1	X_2	Objective fn.
Z_1^*	2.24	0.0	80.6	2.24	0.0	80.6
Z_2^*	0.0	1.87	125.7	0.0	1.88	126.3
Z_3^*	1.63	1.28	144.7	1.74	1.24	146.0

Table 2. Results for the ten beam grillage (stress constraints only)

α	X_1	X_2	X_3	X_4	X_5	X_6	Weight present	1^TM_m present	1^TM_m Ref. 4	Optimal topology present	Optimal topology Ref. 4
1.0	0.0	0.0	1.2	1.2	1.2	1.2	336.0	490.0	490.0		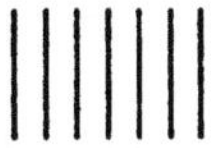
1.2	0.4	0.5	0.0	0.6	3.0	0.2	419.2	570.5	582.7	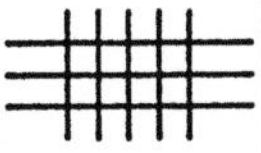	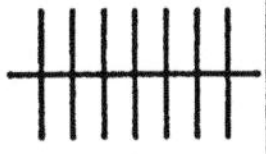
1.5	0.6	0.5	0.0	0.7	1.6	3.1	488.4	623.6	688.3	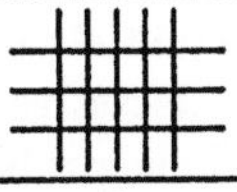	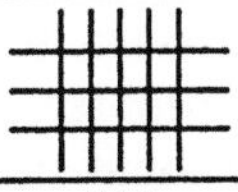
2.0	1.0	2.4	0.0	0.0	0.0	4.5	503.6	725.6	710.0		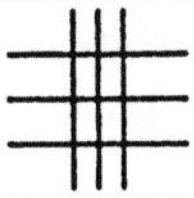
2.3	3.2	3.2	0.0	0.0	0.0	0.0	506.4	652.2	652.2		
2.5	3.1	3.1	0.0	0.0	0.0	0.0	470.5	600.0	600.0		

Table 3. Results for the ten beam grillage (stress and displacement constraints)

α	X_1	X_2	X_3	X_4	X_5	X_6	Weight	Increment in weight	Optimal topology stress & disp. constraints	Optimal topology stress constraints
1.0	0.0	0.0	1.3	1.3	1.3	1.3	364.0	+28.0		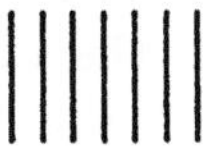
1.5	0.5	0.8	0.0	0.0	4.4	3.1	700.6	+212.2	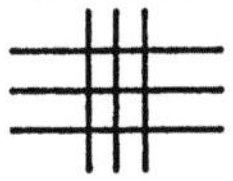	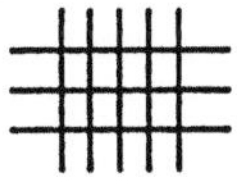
2.0	0.5	4.4	0.0	0.1	0.7	11.5	1047.0	+543.4	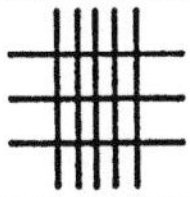	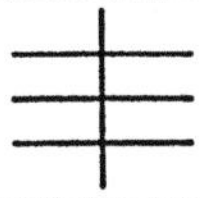
2.5	0.5	11.2	0.1	0.0	0.7	0.5	750.1	+279.6	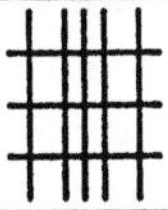	

Optimization of a Linear Objective Function with Logical Constraints

Jacek F. Mączyński

Institute of Fundamental Technological Research, Warszawa

Abstract. When structural safety is assimilated by *Inclusion Principle* to a problem in discretized topology constraints of logical type may be taken account of. This problem is assimilated at least in part to a problem in Design Automation. This in turn is reduced to that of finding a path on a weighted lattice.

1   Preliminaries

The condition of structural safety or to be explicit, the safety of the building, the bridge, the aircraft, the automobile is usually expressed by referring to abstract topology which in soft parlance is: we have to abide within safety limits! Any load on a structure produces a field of stress in it and stress may have 6 components. Sometimes 3 suffice either in a planar case in a case of a homogeneous material. Thus, in any physical, or E^3 , point of the structure there is a mapping from the global load space into the local stress space. When various static or quasistatic loads are applied to the structure within a period of time (0, T), the said mapping produces various points in the stress space of a structure point which may be referred to as physical and local. Thus, a history of loads is mapped onto a swarm of points in the stress space of a physical point. Since the structure is visualized as a continuous body there is a continuum of such swarms. It is the duty of the designer to ensure that the structure will not be damaged or collapse under the effect of any state of stress that may occur in any of the swarms at any point of the structure. This means that we have to find bounds of all swarms and these bounds must abide within the safety bounds.

Since consideration of the whole continuum of physical points of a structure is rarely feasible (unless the structure is suitably parametrized) only a discrete set of physical control points is usually selected to represent a structure in a numerical application.

Safety bounds are given by limit surfaces in the stress space. Some people doubt the existence of distinct limit surfaces and speak about regions in stress space where disastrous damage happens to the material and other regions where the damage is slower to incur and a safe lifetime may be assumed to exist. These latter regions are those of *admissible stress.*

There are theories which suggest unbounded (or partly unbounded) admissible stress, or safe, regions, but they do not seem to be physically justified.

Swarms either discrete or even continuous may be represented by their *least convex hull.*

The vital question of safety, or safety over a lifetime, may thus be formulated briefly in terms of an *Inclusion Principle* in the stress space, viz.:

The convex hull of all the local swarms of points in the stress space must be totally included within the region of admissible stress which is a material characteristic, it is a convex region and it may, partially at least, be reproduced from sample destruction tests in a laboratory.

This Inclusion Principle may serve as a common notion for extensive investigations carried for a very long time and Tresca, Huber, Mises and Hencky are among the earliest in the field.

This Principle is in the spirit of Hausdorff since it emphasizes topological aspects such as inclusion and convexity.

Again, in order to obtain a numerical algorithm, the convex hull of a swarm of points in the stress space may be replaced without loss to safety by a circumscribed hyperpolyhedron and similarly the convex admissible region may be assimilated to an inscribed hyperpolyhedron. Both polyhedra are convex and this is an elementary consequence of the hull and the admissible region being convex.

As a direct consequence of the convex hyperpolyhedra being intersections of halfhyperspaces, a formulation in terms of linear inequalities is readily obtained. Obviously we are looking here for a sufficient condition of safety and not for a necessary one.

Imposing an extremum condition (or at times a minimum such as least overall weight, called dead load, of the structure) a problem in the spirit of Kantorovitch is obtained although not necessarily in the spirit of van Dantzig. This latter remark is due to the relative number of the constraints and the number of unknowns. There are usually many more constraints that there are unknowns. This implies that conventional linear programming simplex methods may prove unsuitable.

It is known that the Inclusion Principle may, at times, under other names be translated into other formulations including those based on penalty functions, augmented Lagrangian or complementarity, nevertheless the boundary of the admissible region being somewhat vaguely defined, a somewhat cruder approach is fully justified if safety is preserved and computation efficiency is enhanced thereby.

Thus an affinity is demonstrated to exist between the problem of structural safety and that of optimisation under linear constraints or inequalities.

Safety requirements demand considering all the relevant cases of loading. This is facilitated when each element of the swarm is a linear combination of some load images as elements of a basis of a linear space because loads themselves are linear combinations of elementary loads.

Again, some combinations of loads are bound to occur and other combinations are highly improbable or excluded by rules of past experience accumulated in Codes of Practice.

To be specific, a nuclear power station may have to withstand its own dead load, high winds and earthquakes. The *dead* load is always present but high wind and an earthquake have a very low probability of occurring simultaneously. This is only an example and we are not concerned here with setting up regulations of a Code of Practice but, on the contrary, our main concern is the use of the existing Codes in engineering computations.

A list of constraints may be much longer in engineering practice. Occurrence of load events may be represented by a set of rules stating mutual exclusion or simultaneous occurrence, rather than by say, probabilities which are, due to our short experience with some structure in question, vastly unknown.

Such rules are written as verbal propositions and thus we arrive into the realm of Wittgenstein and his frequently quoted statement: "Alles was sich überhaupt sagen lässt, lässt sich klar sagen....". It is our objective to explain such rules to a computer and it takes some extra clarity to achieve this.

2 Logical constraints, statement of the problem.

After we have gone through the heuresis outlined in Sec.1. we may be convinced that some problems in design optimisation are quite naturally related to finding an optimum of an objective function f with imposed constraints which are logical rather than algebraic or functional in nature (cf.also [5]). Usually the objective function is a measure of some overall cost but sometimes it is a physical parameter, such as stress, whose extreme values are required.

The constitutive components of a complex piece of equipment are to be found in various "a priori admissible" sets such as catalogues, pricelists &c. The unavoidable selection or decision process is not solely based on individual merits of the items concerned but it is to a great extent concerned with some *compatibility conditions*. As we said, this by far is not the only possible application but a frequent and seemingly illustrative one. The optimisation problem may be stated as follows:

(i) Minimize (or by sign change, maximize):

$$f = \sum_{i=1}^{N} \lambda_i x_i \tag{1}$$

where the parameters λ_i (or "prices") are not known beforehand and x_i are some "intensities " particular to the problem.

(ii) The parameters λ_i of the objective function f assume values from discrete *a priori admissible* sets ("catalogues"):

$$\lambda_i \in \Lambda_i, \quad i \in \{1,...,N\} \tag{2}$$

When necessary, let us introduce subsets ("subcatalogues") such that they exaust the catalogues Λ_i, thus:

$$\Lambda_i^{\alpha_i} \subset \Lambda_i, \ \{\alpha_i \in \{1,\ldots,\alpha_{i_N}\}\},$$

$$\bigcup_{\alpha_i\,|\,1}^{\alpha_{i_N}} \Lambda_i^{\alpha_i} = \Lambda_i \ \{i \in \{1,\ldots,N\}\} \tag{3}$$

The intersections of the subcatalogues may be void but not necessarily so.

(iii) Introducing propositions (cf. [5] and [7]) of the form:

$$S_i^{\alpha_i} = (\lambda_i \in \Lambda_i^{\alpha_i}) \quad where \quad \Lambda_i^{\alpha_i} \subset \Lambda_i, \tag{4}$$

we may relate such propositions by binary (or two-argument) relations (from a relation set $\mathcal{R}$):

$$S_k^{\alpha_k} \ \mathcal{R}_{k\ j}^{\alpha_k \alpha_j} \ S_j^{\alpha_j} \tag{5}$$

where

$$\mathcal{R}_{k\ j}^{\alpha_k \alpha_j} \in \mathcal{R} \tag{6}$$

The set $\mathcal{R}$ of binary relations is isomorphic with some subset of $\mathcal{K}^2$ or Cartesian product of the union of statements:

$$\mathcal{K} = \bigcup_{i\,|\,1}^{N} \ \bigcup_{\alpha_i\,|\,1}^{\alpha_{i_N}} \ S_i^{\alpha_i} \tag{7}$$

since there is a one-to-one mapping of $\mathcal{R}$ onto a subset of $\mathcal{K}^2$.

The logical binary relations between statements of belonging to subcatalogues have usually a verbal form stating some natural compatibility rule of pieces of equipment or of rules of the relevant Code of Practice, such as implication and mutual exclusion.

To complete this set of notions we bear in mind that a design $\mathcal{D}$ is a set composed of individual items:

$$\mathcal{D} = \bigcup_{i\,|\,1}^{N} \lambda_i, \ \{\lambda_i \in \Lambda_i\} \tag{8}$$

for all $i \in \{1,\ldots,N\}$

The set of all designs ("good" and "bad") is as rich as the Cartesian product:

$$\Lambda_1 \times \Lambda_2 \times \cdots \times \Lambda_N \tag{9}$$

Obviously a vast majority of the elements of this set have to be rejected on grounds of compatibility and optimality.

3 Binary relations and tautologies

Implication constraints may be written in the form

$$S_i^{\alpha_i} \rightarrow S_k^{\alpha_k} \tag{10}$$

where $\rightarrow$ denotes the logical operator of implication with the usual logical values. Since we expect that λ's belong to some catalogue we write the tautology of counterposition in the form:

$$[(\lambda_i \in \Lambda_i^{\alpha_i}) \rightarrow (\lambda_j \in \Lambda_j^{\alpha_j})]$$

$$\leftrightarrow \tag{11}$$

$$[(\lambda_j \in \Lambda_j \setminus \Lambda_j^{\alpha_j}) \rightarrow (\lambda_i \in \Lambda_i \setminus \Lambda_i^{\alpha_i})]$$

Some further tautologies are useful, e.g.:

$$[(\lambda_i \in \Lambda_i^{\alpha_i}) \ NAND \ (\lambda_j \in \Lambda_j^{\alpha_j})]$$

$$\leftrightarrow$$

$$NOT\,[(\lambda_j \in \Lambda_j) \ AND \ (\lambda_i \in \Lambda_i)]$$

$$\leftrightarrow \tag{12}$$

$$[NOT(\lambda_i \in \Lambda_i^{\alpha_i}) \ OR \ NOT(\lambda_j \in \Lambda_j^{\alpha_j})]$$

$$\leftrightarrow$$

$$(\lambda_i \in \Lambda_i)] \rightarrow (\lambda_j \in \Lambda_j \setminus \Lambda_j^{\alpha_j})]$$

Here NOT denotes negation, AND conjunction, $NAND$ exclusion, OR alternative and $\leftrightarrow$ equivalence. We have $\Lambda \setminus \Lambda = \emptyset$ for any Λ.

Manipulation of the logical constraints consists in first, writing them down as statements and relations and next some ordering in the way the λ's are numbered. This is made easier by the tautologies such as Eq.(11) or Eq.(12). Next a graphic representation is used to visualize the constraints. Implication appears in the form of Fig.1.

Semi-thick arrows intend to emphasize that implications refer to the elements of the sets and stand for expanding implication fans which would

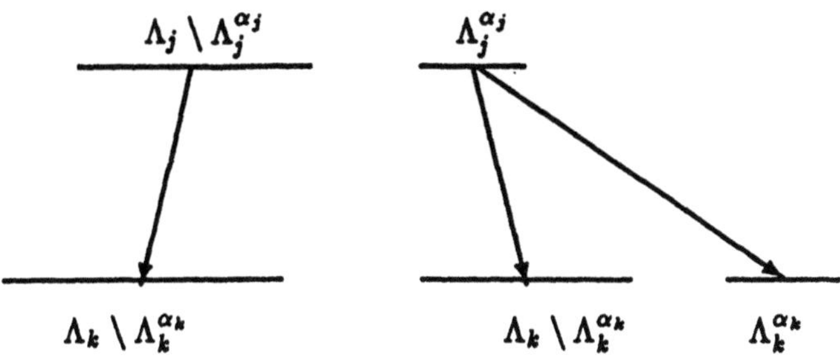

Figure 1: Graph edge representation of logical constraints

otherwise be confusing. By linking together graphs of the type shown in Fig.1 a (prohibitively large) tree structure would be arrived at. This would show all the feasible designs. Any such design corresponds to a path from the first element to the last over the successive stages of ramifications.

Fortunately the tree obtained is an *avalanche* (cf.[6]) of a directed network type graph which is assimilable with a *lattice* (cf.[3] and below).

The actual construction of such a directed network requires some permutations since implications have to be written so as to have left hand side λ's precede the λ's on the right-hand side on the list. Counterposition is used for this purpose. Another obvious economy rule demands that the relations should link λ's as close to each other on the list as possible.

After this ordering has been performed the implication chain is checked for such features as subcatalogues overlapping or whether the chain is not a simple sum of disjoint chains.

4 Ordering the set $\mathcal{K}$.

The steps outlined above are best justified by discussing the set in more detail. The relation set $\mathcal{R}$ is endowed with reflexivity ($\mathcal{S} \rightarrow \mathcal{S}$) and transitivity. This latter property is a direct result of the ordering of the λ's. The manipulation consists in:

- (i) enlarging the set $\mathcal{K}$ to comprise the least (L) and greatest (G) element,

- (ii) when $j = i + 1$ in Eq.(10) no enlarging at level j is necessary,

- (iii) when $j > i + 1$ additional intermediate levels are introduced so as to make rule (ii) apply.

The lattice is a representation of the set of propositions $\mathcal{K}$ and contains ordered chains of arcs running from L to G.

In practical applications ordered chains of arcs are called paths, a lattice is conveniently called a network [3] and for the purpose of optimisation the paths have computable lengths since weights are ascribed to arcs.

5 Optimisation on a lattice.

Any chain on the lattice corresponds to a class of admissible designs since compatibility is ensured, the logical constraints having been used to construct the lattice.

Optimisation ascribes a weight or an intensity to every arc which is incident to a statement S. When minimizing, the weight is given a lowest value from the subcatalogue $\Lambda_i^{\alpha_i}$. Conversely the largest value is taken when maximizing. Thus no more than two elements from the subcatalogues are considered. Sublattices may be eliminated at an early stage. The cost of the whole design arises as the sum over a path in the lattice.

The automation occurs as yet when the lattice has been fully organized and then the minimum (resp. maximum) path length is found by a version of path optimisation in a network. An illustrative example may be found in [2].

References

[1].Arnold, B.H., 1963: Logic and Boolean Algebra, Prentice-Hall.

[2].Eimer, Cz.; Mączyński J., 1976: On optimal shell prestressing, *J. Struct. Mech.* 4, 3, 289-305.

[3].Ford, L.R. Jr.; Fulkerson, D.R., 1962: Flows in networks, Princeton.

[4].Grzegorczyk, A., 1973: An outline of mathematical logic, Reidel-PWN Warszawa.

[5].Hammer, P.L., 1979: Boolean elements in combinatorial optimisation, *Ann. of Discr. Math.* 4, pp. 51-71.

[6].Lipski, W.; Marek W. 1973: On Hamiltonian paths in a graph, CC PASc Reports, no. 130, Warszawa.

[7].Pogorzelski, W.A., 1973: The classical propositional calculus (in Polish), PWN Warszawa.

On Optimal Structural Segmentation Problem

K. Dems[1]   and   Z. Mróz[2]

[1] Łódź Technical University, Łódź, Poland
[2] Institute of Fundamental Technological Research,
Warsaw, Poland

ABSTRACT *Large structures are usually composed of elements assembled on the construction site by properly designed connections. The problem arises to provide optimal size and number of elements provided the global structure dimensions and loading are specified. The problem is formulated by assuming the element cost to be a non-linear function of its size and cost of connection forced. The number of elements, element sizes and their layout now constitute the design parameters. A two-level procedure is developed which allows to determine number of elements and optimal segmentation for beam, plate or truss structures. Some illustrative examples are treated in details.*

1. INTRODUCTION

The present paper is concerned with an optimal design problem for which the number of structural elements and

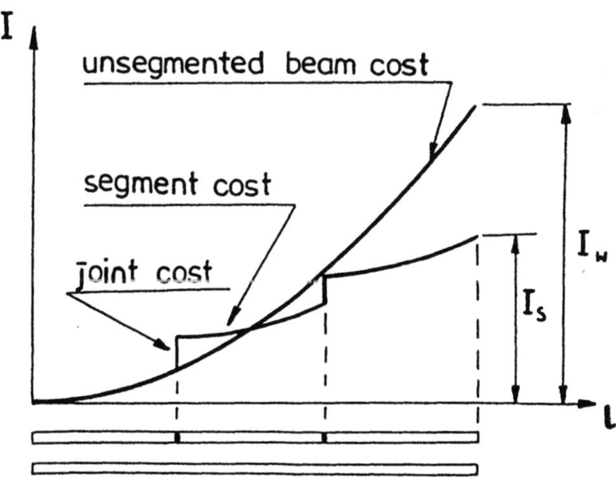

Figure 1. Cost of segmented and unsegmented structure

their size are to be determined besides the usual dimensional or shape variables. As large structures are compo-

48

sed of elements interconnected at joints, their number
can be varied provided technological and service constraints are satisfied. The cost function is assumed to increase non-linearly. with the element size, thus representing not only the material cost, but also the cost of
transportation, manufacture and assemblage. The connection cost is assumed either to be constant or to depend on
generalized forces transmitted by the connections. The
mean idea of the above formulated problem is shown in
Fig. 1. It is observed that for increasing size of element the cost of structure is increased very fast in comparison to segmented structure with joints for which the
total cost can be reasonably reduced. Though the number
of elements is a discrete design parameter, in the analysis it will be treated as a continuous function. A two-
level procedure developed in this paper provides a systematic approach in specifying the number of elements and
optimal segmentation for beam, disk and truss structures.
Some examples provide the illustration of optimization
procedure.

2. BEAM AND FRAME SEGMENTATION PROBLEM

Let us start from a simple case, namely a beam structure with specified support and loading conditions, Fig.
2. Assume the beam to be divided into p segments connected at p-1 joints. As the total beam length L is specified, there is

$$\sum_{i=1}^{p} l_i - L = 0 \tag{1}$$

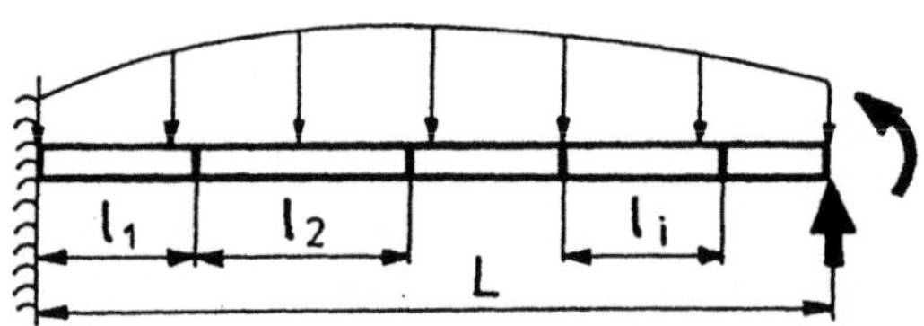

Figure 2. Segmented beam structure

The cost function of the i-th segment is assumed in the
form

$$s_i = s_o + s_1\left(\frac{l_i}{l_o}\right) + s_m\left(\frac{l_i}{l_o}\right)^m \qquad i=1,2,\ldots,p \tag{2}$$

where l_o is the characteristic segment length, s_o, s_1, s_m
are the segment cost coefficients and m denotes the cost

exponent. The coefficients s_1 and s_m are assumed to depend on cross-section parameters.

In particular, when $s_1 = A_i l_o$, $s_o = 0$, $s_m = 0$, where A_i is the cross-section area, then $s_i = A_i l_i$ represents the segment material volume. The optimal segmentation problem, for which material volume was minimized subject to specific constraint on elastic compliance or stress was already considered by Masur [1,2] and Dems and Mróz [7,9]. The slightly similar problem was also considered by Rozvany and Mróz [3,4], Szeląg and Mróz [5] and Garstecki and Mróz [8]. Prager [6] discussed the optimal layout of truss with finite number of joints.

The cost of joint connecting two segments is now assumed in the form

$$c_k = c_o + c_1\left(\frac{M_k}{M_o}\right) + c_n\left(\frac{M_k}{M_o}\right)^n \qquad k=1,2,\ldots,p-1 \qquad (3)$$

where c_o, c_1, c_n are joint cost coefficients, M_k is the bending moment at k-th joint, M_o denotes the reference moment and n is the joint cost exponent. The total cost of the segmented beam is

$$I = \sum_{i=1}^{p} s_i + \sum_{k=1}^{p-1} c_k \qquad (4)$$

The optimal design problem for minimum compliance will now be formulated as follows

$$\text{Minimize} \quad I$$
$$\text{subject to} \quad \sum_{i=1}^{p} l_i - L = 0 \qquad (5)$$
$$C = \int_o^L W(M, A_i, l_i)\, dx \leq C_o$$

where C denotes the global structural compliance measured by the complementary energy of the beam. In writing (5) it was assumed that the global compliance does not depend on connection stiffness but only on cross-sectional stiffness and length of structure members.

In order to obtain as insight into the problem, let us assume for the moment that the mean compliance constraint does not affect the beam segmentation. This assumption is valid when the design is uniform with all cross-sectional parameters constant along the beam. Then the variation of complementary energy equals

$$\delta C = \int_o^L \frac{\partial W}{\partial M}\, \delta M\, dx = \int_o^L x\, \delta M\, dx = 0 \qquad (6)$$

50

for any statically admissible variation δM. To obtain an analytical solution, assume uniform segmentation and the connection cost as constant, thus

$$l_i = \frac{L}{p} \qquad i=1,2,\ldots,p \quad , \qquad c_k = c_o \qquad k=1,2,\ldots p-1$$

$$s_i = s_o + s_1\left(\frac{L}{pl_o}\right) + s_m\left(\frac{L}{pl_o}\right)^m \qquad i=1,\ldots p \tag{7}$$

where p denotes the number of segments. The total cost (4) now equals

$$I = ps_o + ps_1\left(\frac{L}{pl_o}\right) + ps_m\left(\frac{L}{pl_o}\right)^m + (p-1)c_o \tag{8}$$

Introducing non-dimensional quantities

$$\beta_1 = \frac{s_1}{s_o} \quad , \quad \beta_m = \frac{s_m}{s_o} \quad , \quad \alpha_o = \frac{c_o}{s_o} \quad , \quad \hat{I} = \frac{I}{s_o} \tag{9}$$

we have

$$\hat{I}(p) = p\left[1 + \beta_1\left(\frac{L}{pl_o}\right) + \beta_m\left(\frac{L}{pl_o}\right)^m\right] + (p-1)\alpha_o \tag{10}$$

In order to minimize $\hat{I}(p)$, let us assume that $p \in [0,\infty]$ is a continuous variable and differentiate (10) with respect to p. Then we obtain the optimality condition in the form

$$\frac{d\hat{I}}{dp} = 1 + \alpha_o - (m - 1)\left(\frac{L}{pl_o}\right)^m \beta_m = 0 \tag{11}$$

The stationarity conditions (11) now provides the value of p

$$p = \frac{L}{l_o}\left(\frac{m-1}{1+\alpha_o}\beta_m\right)^{\frac{1}{m}} \tag{12}$$

corresponding to a strong minimum of $\hat{I}(p)$. We have thus obtained an analytical assessment of the optimal number of segments. After selection of p as a closest integer to (12), we may pass to the standard optimal design of a beam by determining A_i from the optimality conditions.

The assumption (7) can now be generalized by accounting for varying connection cost with the bending moments transferred by the connections. Assume instead of (7) that for the uniform segmentation, $l_i = L/p$, we have

$$c_k = c_o + c_1\left(\frac{M_k}{M_o}\right) + c_n\left(\frac{M_k}{M_o}\right)^n \qquad k=1,2,\ldots,p-1$$

$$s_i = s_o + s_1\left(\frac{L}{pl_o}\right) + s_m\left(\frac{L}{pl_o}\right)^m \qquad i=1,..p \qquad (13)$$

The non-dimensional cost function (10) now equals

$$\hat{I}(p) = p[1 + \beta_1\left(\frac{L}{pl_o}\right) + \beta_m\left(\frac{L}{pl_o}\right)^m] + (p-1)\alpha_o +$$
$$\sum_{k=1}^{p-1}\left[\alpha_1\left(\frac{M_k}{M_o}\right) + \alpha_n\left(\frac{M_k}{M_o}\right)^n\right] \qquad (14)$$

where $\alpha_1 = c_1/s_o$ and $\alpha_n = c_n/s_o$. The stationarity condition for $I(p)$ can now be generated similarly as previously. However, the analytical solution for p can not be obtained in a closed form.

A more general formulation can be proposed by assuming unspecified segment lengths l_i for a given number of segments p. Instead of (14) we now have

$$\hat{I}(l_i) = \sum_{i=1}^{p} [1 + \beta_1\left(\frac{l_i}{l_o}\right) + \beta_m\left(\frac{l_i}{l_o}\right)^m] + (p-1)\alpha_o +$$
$$\sum_{k=1}^{p-1}\left[\alpha_1\left(\frac{M_k}{M_o}\right) + \alpha_n\left(\frac{M_k}{M_o}\right)^n\right] \qquad (15)$$

subject to

$$\sum_{i=1}^{p} l_i - L = 0 \qquad (16)$$

The augmented cost function now is

$$\hat{I}^a(l_j,\lambda) = \hat{I} + \lambda(\sum_{i=1}^{p} l_i - L) \qquad (17)$$

and the optimality conditions

$$\frac{\partial \hat{I}^a}{\partial l_j} = 0 \qquad j=1,2,...,p \quad , \quad \frac{\partial \hat{I}^a}{\partial \lambda} = 0 \qquad (18)$$

provide the set of non-linear algebraic equations for l_i and the Lagrange multiplier λ. The optimality conditions now take the form

52

$$\left[\beta_1 + m\beta_m\left(\frac{l_i}{l_o}\right)^{m-1}\right]\frac{1}{l_o} + \sum_{i=1}^{p-1}\left[\alpha_1 + n\alpha_n\left(\frac{M_i}{M_o}\right)^{n-1}\right]\frac{Q_i}{M_o} + 1 = 0 \qquad (19)$$

$$\sum_{i=1}^{p} l_i - L = 0$$

where

$$Q_i = \frac{dM_i}{dl_i} = \left(\frac{dM}{dx}\right)_{x_i=\sum\limits_{j=1}^{i} l_j} \qquad (20)$$

is the shear force transferred at the connection between the segments i and i+1.

Consider now a more general case when the compliance of connections is included into the formulation. Assume that each connection now constitutes an elastic hinge with the discontinuity in deflection slope $[\theta] = \theta^+ - \theta^-$ related to connection stiffness D by the relation $M_i = D[\theta]_i$, so that the connection stress energy now is

$$\Pi_M^j = \frac{1}{2} M_i[\theta]_i = \frac{1}{2}\frac{M_i^2}{D} = \frac{1}{2EI}\frac{M_i^2}{d} \qquad (21)$$

where EI denotes the beam bending stiffness and d = D/EI is the relative connection stiffness. The total connection stress energy is

$$\Pi_M^c = \sum_{i=1}^{p-1} \Pi_M^i = \frac{1}{2EI}\sum_{i=1}^{p-1}\frac{M_i^2}{d} \qquad (22)$$

and the total complementary beam energy equals

$$\Pi_M = \Pi_M^b + \Pi_M^c = \frac{1}{2EI}\left[\sum_{i=1}^{p}\int_o^{l_i} M^2 dx + \sum_{i=1}^{p-1}\frac{M_i^2}{d}\right] \qquad (23)$$

Let us now assume that the beam cross-sectional area A is related to cross-sectional area A_o in reference configuration and the cross-sectional moment of inertia is proportional to a square of cross-sectional area A. Thus, we can write

$$A = \mu A_o \quad , \quad I = kA^2 = k(\mu A_o)^2 = \mu^2 I_o \quad , \quad d = \frac{d_o}{\mu^2} \qquad (24)$$

where μ can be treated as additional design parameter and d is a relative connection stiffness in reference configuration. In view of (24), the total complementary energy (23) can be rewritten in the form

$$\Pi_M = \frac{1}{2EI_o} \left\{ \frac{1}{\mu^2} \left[\sum_{i=1}^{p} \int_{o}^{l_i} M^2 dx + \sum_{i=1}^{p-1} \frac{M_i^2}{d} \right] \right\} \qquad (25)$$

We should note now that the segmentation affects the mean compliance of beam through varying number of elastic joints and varying beam cross-section. Then the mean compliance constraint can be written now in the form

$$\Pi_M - \Pi_M^o \leq 0 \qquad (26)$$

where Π_M^o denotes the total complementary energy of unsegmented beam in reference configuration and is expressed by

$$\Pi_M^o = \frac{1}{2EI_o} \int_{o}^{L} M^2 dx \qquad (27)$$

Introducing the non-dimensional cost coefficients

$$\alpha_o = \frac{c_o}{A_o s_o} \quad , \quad \alpha_1 = \frac{c_1}{A_o s_o} \quad , \quad \alpha_n = \frac{c_n}{A_o s_o} \qquad (28)$$

the non-dimensional cost function of segmented beam can be expressed in the form similar to (15), namely

$$\hat{I}(l_i,\mu) = \frac{I}{A_o s_o} = \sum_{i=1}^{p} \mu [1 + \beta_1 \left(\frac{l_i}{l_o} \right) + \beta_m \left(\frac{l_i}{l_o} \right)^m] + (p-1)\alpha_o +$$

$$\sum_{k=1}^{p-1} \left[\alpha_1 \left(\frac{M_k}{M_o} \right) + \alpha_n \left(\frac{M_k}{M_o} \right)^n \right] \qquad (29)$$

The optimization problem can now be stated in the form

$$\text{Minimize} \quad \hat{I}(l_i,\mu)$$

$$\text{subject to} \quad \sum_{i=1}^{p} l_i - L = 0 \qquad (30)$$

$$e_M - e_M^o = 0$$

54

where

$$e_M = 2EI_o\Pi_M \quad , \quad e_M^O = 2EI_o\Pi_M^O \tag{31}$$

The augmented cost function can now be written in the form

$$\hat{I}^a = \hat{I} + \lambda_1 \left(\sum_{i=1}^{p} l_i - L \right) + \lambda_2 (e_M - e_M^O) \tag{32}$$

where λ_1 and λ_2 are the Lagrange multipliers. Thus, the optimality conditions following from the stationarity of functional (32)

$$\frac{\partial \hat{I}}{\partial l_i} + \lambda_1 + \lambda_2 \frac{\partial e_M}{\partial l_i} = 0 \quad , \quad \frac{\partial \hat{I}}{\partial \mu} + \lambda_2 \frac{\partial e_M}{\partial \mu} = 0$$

$$\sum_{i=1}^{p} l_i - L = 0 \quad , \quad e_M - e_M^O = 0 \tag{33}$$

provide the set of non-linear algebraic equations for l_i, μ and Lagrange multipliers λ_1, λ_2.

3. EXAMPLES OF OPTIMAL BEAM SEGMENTATION

Let be given a beam of length L and uniform cross-section A shown in Fig. 3. The beam is clamped on the left end, supported on the right end and subjected to uniform load t. The beam is divided into p segments of length l_i connected at p-1 joints.

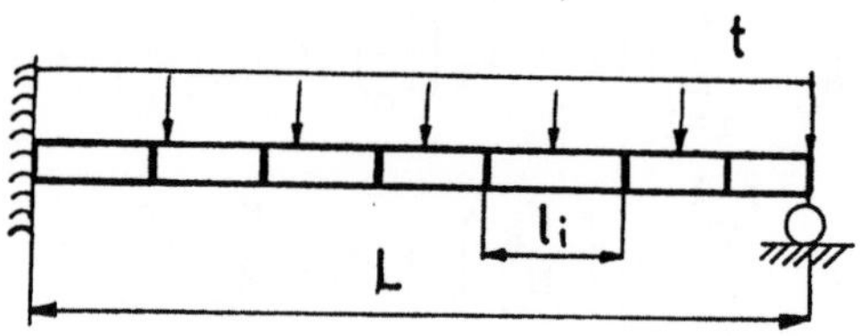

Figure 3. Uniformly loaded segmented beam

Consider first the case of uniform segmentation and constant joint cost in which the optimal number of segments will be derived. The optimal values of p follow then from (12) and are given in Tab.1 for some prescribed values of $\beta_m = 2.0$ and $L/l_o = 4$. In Figure 4 the distri-

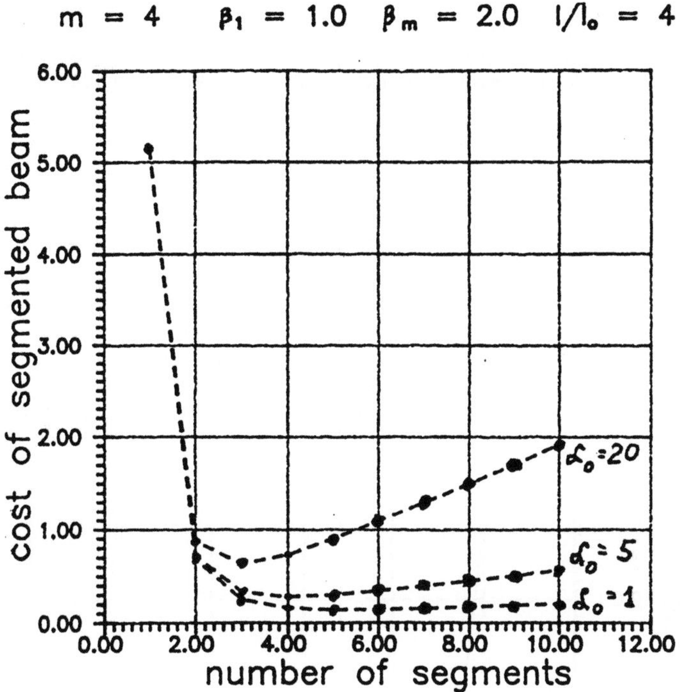

Figure 4. Cost of beam versus number of segments

bution of beam cost versus varying number of segments is plotted for some prescribed values of m, β_1, β_m, L/l_o and

Table 1. Values of p for uniform design

α_o	1.00	5.00	20.00	100.00
m	number	of	segments	
2	4	2	1	1
3	5	3	2	1
4	5	4	3	2
5	5	4	3	2
6	5	4	4	3
7	5	4	4	3
8	5	4	4	3

α_o. It is easy to observe that the optimal number of segments corresponds to that shown in Table 1.

As the second example consider the optimization problem (30) for the beam from Figure 3. Thus we have now a beam composed from initially given number of segments of unspecified in advance optimal lengths and cross-sectional area, connected with elastic joints of stiffness D. The optimal values of segment lengths l_i and cross-sectional area coefficient μ are derived from optimality conditions (33). The results of calculations for loading intensity t = 10, exponents m = 4 and n = 2, cost coefficients $\beta_1 = 1$, $\beta_m = 2$, $\alpha_o = \alpha_1 = 0$, $\alpha_n = 20$, relative joint stiffness $d_o = 100$ and L = 1, $l_o = 0.25$ are presented in Figure 5. The curve 'initial' corresponds to the initial beam composed from p segments of uniform length, whereas the 'optimal' curve corresponds to the beam with optimal segment lengths. As it can be observed, for some initial numbers of segments, its optimal number (and also the number of joints) is reduced due to the fact that the

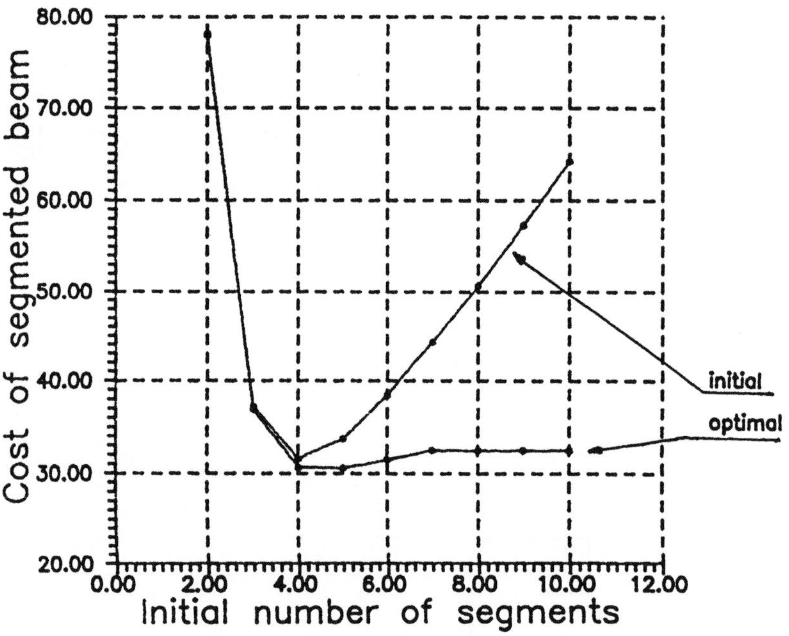

Figure 5. Initial and optimal beam cost

length of some segments was tending to zero during the optimization process. In Table 2, the initial and optimal lengths of segments are presented for initial number of segments equals to 8. This number was reduced to 7 during the optimization process. The corresponding cross-sectional area coefficient μ took the optimal value 1.012, and the decreasing of total beam cost in comparison to initial design was about 36%.

Table 2. Optimal segment lengths

Segment	Segment length	
number	initial	optimal
1	0.125	0.221
2	0.125	0.034
3	0.125	0.012
4	0.125	0.147
5	0.125	0.283
6	0.125	0.238
7	0.125	0.065
8	0.125	-

4. OPTIMAL DISK SEGMENTATION

In this Section the problem of disk segmentation will be outlined using as an example a rectangular disk of dimensions a and b, subjected to uniform load t on upper edge, Fig. 6. The disk is divided into rectangular segments by p-1 vertical and q-1 horizontal connection lines, so that the dimensions of typical disk segment are a_i and b_j, i=1,..,p, j=1,..q.

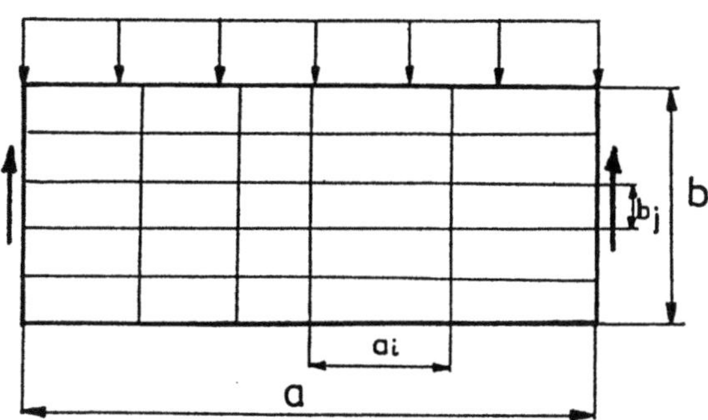

Figure 6. Segmented disk

The cost of segments and connection lines can be assumed in the form similar to (13) for segmented beam. Thus, the non-dimensional segment cost is assumed in the form

$$\hat{s}_{ij} = 1 + \beta_1 \left(\frac{a_i b_i}{a_o b_o}\right) + \beta_m \left(\frac{a_i b_i}{a_o b_o}\right)^m \qquad \begin{array}{l} i=1,..,p \\ j=1,..,q \end{array} \qquad (34)$$

where a_o and b_o denote the characteristic segment dimensions and β_1 and β_2 are the cost coefficients. Similarly, the cost of horizontal and vertical connection lines are

58

expressed by

$$\hat{c}_j^h = \int_{-a/2}^{a/2} \left[\alpha_o + \alpha_1 \left(\frac{\sigma_r(x,y_j)}{\sigma_o} \right) + \alpha_n \left(\frac{\sigma_r(x,y_j)}{\sigma_o} \right)^n \right] dx$$

$$\hat{c}_i^v = \int_{-b/2}^{b/2} \left[\alpha_o + \alpha_1 \left(\frac{\sigma_r(x_i,y)}{\sigma_o} \right) + \alpha_n \left(\frac{\sigma_r(x_i,y)}{\sigma_o} \right)^n \right] dy$$

$$(35)$$

where α_o, α_1, α_n denote the connection line cost coefficients, σ_o is the reference stress and reduced stress σ_r
equals

$$\sigma_r = \sqrt{\sigma^2 + 3\tau^2} \tag{36}$$

Here σ and τ denote the normal and tangential stress components along connection line.

The total non-dimensional cost of segmented disk can
now be expressed in the form

$$\hat{I} = \sum_{i=1}^{p} \sum_{j=1}^{q} \hat{s}_{ij} + \sum_{j=1}^{q-1} \hat{c}_j^h + \sum_{i=1}^{p-1} \hat{c}_i^v \tag{37}$$

and the following optimization problem can be stated

Minimize $\quad \hat{I}$

$$\text{subject to} \quad \sum_{i=1}^{p} a_i - a = 0 \quad , \quad \sum_{j=1}^{q} b_j - b = 0 \tag{38}$$

The optimality conditions for the problem (38) follow
from stationarity of augmented cost functional.

Let consider as the first case the uniform segmentation of a disk in x and y directions and constant connection line cost. Thus we have

$$a_i = \frac{a}{p} \quad , \quad b_j = \frac{b}{q} \quad , \quad \alpha_1 = \alpha_n = 0 \tag{39}$$

and the total cost (37) takes the form

$$\hat{I} = pq + \beta_1 \left(\frac{ab}{a_o b_o} \right) + \beta_m \left(\frac{ab}{a_o b_o} \right)^m (pq)^{1-m} +$$

$$[(q-1)a + (p-1)b] \tag{40}$$

whereas both geometrical constraints are satisfied identically. Treating the numbers of segmentation p and q as continuous variables, their optimal values can be obtained as the solution of the following set of non-linear equations

$$\frac{\partial \hat{I}}{\partial p} = 0 \quad , \quad \frac{\partial \hat{I}}{\partial q} = 0 \tag{41}$$

Selecting for p and q the closest integer values to solution of (41) we obtain the optimal numbers of segments in x and y directions. An example of obtained results is presented in Table 3.

Table 3. Values of p and q for uniform design
$a=1.0$ $b=0.5$ $a_o=0.25$ $b_o=0.125$ $\beta_1=1.0$ $\beta_2=2.0$

α_o	1.0		2.0		20.0		100.0	
m	optimal segmentation numbers							
	p	q	p	q	p	q	p	q
2	5	4	6	3	5	2	3	1
3	6	4	6	4	5	3	4	2
4	6	4	6	4	6	3	4	3
5	6	4	6	4	6	3	5	3
6	6	4	6	4	6	3	5	3
7	7	3	7	3	6	3	5	3
8	7	3	7	3	6	3	5	3

Assume now that the numbers of segments p and q are specified in advance. Thus the optimization problem (30) is reduced to selecting the optimal values of a_i and b_j. The optimality conditions are then written as

$$\frac{\partial \hat{I}}{\partial a_i} + \lambda_1 = 0 \quad i=1,..,p \quad , \quad \frac{\partial \hat{I}}{\partial b_j} + \lambda_2 = 0 \quad j=1,..,q$$

$$\sum_{i=1}^{p} a_i - a = 0 \quad \sum_{j=1}^{q} b_j - b = 0 \tag{42}$$

and constitute a set of non-linear algebraic equations for a_i, b_j and Lagrange multipliers λ_1, λ_2. The results of calculations for some problem data are presented in Table 4. We should note the reduction of optimal segmentation number in y direction. The calculated decreasing of total optimal cost in comparison to cost of initial design was about 29%.

Table **4**. Optimal segment lengths p=5 q=5 m=4 t=1
β_1=1 β_2=2 α=100 a=1 b=0.5 a_o= 0.25 b_o= 0.125

	i =	1	2	3	4	5
a_i	initial	0.200	0.200	0.200	0.200	0.200
	optimal	0.195	0.199	0.212	0.199	0.195
	j =	1	2	3	4	5
b_j	initial	0.100	0.100	0.100	0.100	0.100
	optimal	0.244	0.223	0.032	0.001	–

REFERENCES

1. Masur, E.F., Optimal structural design for a discrete set of available structural members, Comp. Meth. Appl. Mech. Eng., 3, 1974
2. Masur, E.F., Optimality in presence of discreteness and discontinuity, Proc. IUTAM Symp. on Optimization in Struct. Design, Warsaw 1974, Eds. A. Sawczuk, Z. Mróz, Springer Verl., 441-454,1975
3. Rozvany, G.I.N. and Mróz, Z., Column design: optimization of support condition and segmentation, J. Struct. Mech., 5, 279-290,1977
4. Rozvany, G.I.N. and Mróz, Z., Optimal design taking cost of joints into account, J. Eng. Mech. Div. ASCE,101, 917-921, 1975
5. Szeląg, D. and Mróz, Z., Optimal design of vibrating beams with unspecified support reactions, Comp. Meth. Appl. Mech. Eng., 19, 33-349,1979
6. Prager, W., Optimal layout of trusses with finite number of joints, J. Mech. Phys. Solids, 26, 241-250, 1978
7. Dems, K. and Mróz, Z., Optimal shape design of multicomposite structures, J. Struct. Mech., 8, 309-329,1980
8. Garstecki, A. and Mróz, Z., Optimal design of supports of elastic structures subjected to loads and initial distortions, Mech. Struct. Mach., 8, 47-68,1987
9. Dems, K. and Mróz, Z., Shape sensitivity analysis and optimal design of disks and plates with strong discontinuities of kinematic fields, Int. J. Solids Struct.,29, 437-463,1992

Application of Discrete Optimization Techniques to Optimal Composite Structures

Elke Schäfer[1], Johannes Geilen[2] and Hans A. Eschenauer[3]

[1]Ingenieurbüro Dr. Schäfer, 65594 Runkel, Germany
[2]Dr. Reinold Hagen Stiftung,Technologie-Zentrum Kunststoff, 53229 Bonn, Germany
[3]Forschungszentrum für Multidisziplinäre Analysen und Angewandte Strukturoptimierung
 FOMAAS, Institut für Mechanik und Regelungstechnik, Universität-GH Siegen,
 57076 Siegen, Germany

1. Introduction

In optimizing composite structures particular considerations must be given to the manufacturing possibilities and the distinctive features of the materials applied. To meet all requirements and to gain practicable results for industrial and technical purposes, mixed continuous-discrete optimization problems will often occur.

Usually these problems can be solved by purely continuous optimization algorithms. Difficulties, however, will arise if the design obtained by rounding to discrete values is infeasible or if the objective function is highly sensitive to changes of the discrete design variables or if discrete variables cannot be interpolated. As a consequence, continuous-discrete optimization algorithms have to be employed.

At the *Institut für Mechanik und Regelungstechnik* in Siegen several optimization algorithms have been programmed and applied to solve mixed structural optimization problems.

2. Mathematical Definition

The methods of mathematical programming should be applied early on, during the technical design process of composite structures, because the involved materials offer a lot of producible combinations. First an abstract mathematical formulation of the problem is necessary, one which assumes a quantitative coherence between the objective functions, the technical and physical constraints and the design variables. The following general formulation of a continuous-discrete vector optimization problem covers all special cases, even the scalar continuous problem

$$\min \{ f(x) \} \qquad (1)$$
$$x_c \in X_c$$
$$x_d \in X_d$$

with the continuous and the discrete design spaces X_c and X_d

$$X_c := \{ x_c \in \mathbb{R}^{N_c} \mid x_l \leq x_u \,;\; g(x_c, x_d) \geq 0 \,;\; h(x_c, x_d) = 0 \} \,, \qquad (2)$$

62

$$X_d := \{ \ x_d \in \mathbf{R}^{N_d} \mid x_{dj} \in X_j \ \forall \ j = 1, ..., N_d \ ; \ g(x_c, x_d) \geq 0 \ ; \ h(x_c, x_d) = 0 \ \} \quad (3)$$

and the N_d discrete subsets

$$X_j := \{ \ x_{dj}^{(1)}, x_{dj}^{(2)}, ..., x_{dj}^{(nj)} \ \} \ , \ X_j \subset \mathbf{R}^1 \ \forall \ j = 1, ..., N_d \quad (4)$$

where

$f(x) = (\ f_1(x), ..., f_K(x) \ )^T$	objective function vector,
$x = (\ x_c^T, x_d^T \ )^T$	vector of $(N_c + N_d)$ design variables,
x_c	vector of N_c continuous design variables,
x_d	vector of N_d discrete design variables,
x_l, x_u	vectors of lower and upper bounds of x_c,
$g(x_c, x_d), h(x_c, x_d)$	vectors of inequality and equality constraints,
n_j	number of discrete values of the subset X_j.

Compared with a purely continuous optimization problem, the continuous-discrete optimization problem (1) has some severe drawbacks, e.g. the well-known Kuhn-Tucker optimality criteria cannot be applied. Although the design space of the discrete design variables contains a finite number of values, there exists an infinite number of solutions in the case of vector optimization problems. Additionally, the functional efficient boundary may not be joined together.

Fig. 2.1 shows the influence of continuous, discrete or mixed design variables on the solution of vector optimization problems, by means of the following two-dimensional mathematical example

$$\min \ \{ \ f(x) \} \quad \text{with} \quad f = (\ f_1, f_2 \ )^T , \ g = (\ g_1, g_2 \ )^T \quad (5)$$
$$x_c \in X_c$$
$$x_d \in X_d$$

where
$$f_1 = 2 - x_1, \quad f_2 = x_1 + x_2 - 2x_2 - 1 ,$$
$$g_1 = 4x_1 - x_1^2 + x_2 - 5 \geq 0 , \ g_2 = -x_1 - 2x_2 + 7 \geq 0$$

(a) continuous problem, $N_c=2$:
$$x = x_c = (\ x_1, x_2 \ )^T ,$$
$$x_l = (\ 0, 0 \ )^T , \ x_u = (\ 3.5, 4.0 \ )^T$$

(b) discrete problem, $N_d=2$:
$$x = x_d = (\ x_1, x_2 \ )^T ,$$
$$X_1 := \{ \ 0., 0.3, 0.6, ..., 3.3 \ \}$$
$$X_2 := \{ \ 0., 0.2, 0.4, ..., 4.0 \ \}$$

(c) continuous-discrete problem, $N_c=1$, $N_d=1$:
$$x = (\ x_c^T, x_d^T \ )^T ,$$
$$x_c = (\ x_1 \ ) , \ x_d = (\ x_2 \ ) ,$$
$$x_l = (0) , \ x_u = (3.5) ,$$
$$X_1 := \{ \ 0., 0.2, 0.4, ..., 4.0 \ \}$$

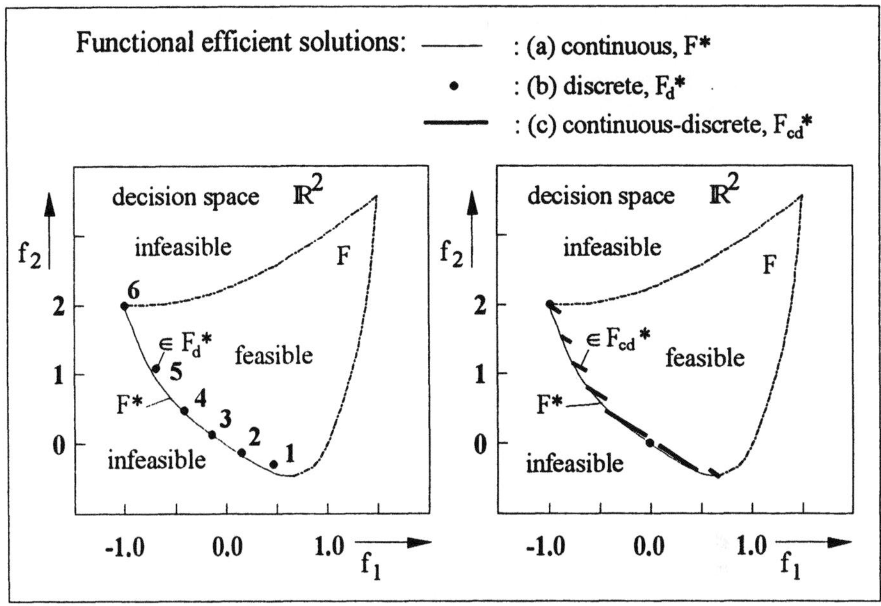

Fig. 2.1. Functional efficient solutions of problem (5) with design sets (a), (b) and (c)

To solve the vector problem (1), it must be reduced to a scalar optimization problem by means of a substitute function, which describes the individual preferences of a decision-maker (see [3]). The substitute problem yields one functional efficient solution $\mathbf{f}^*=(f_1{}^*, f_2{}^*, ..., f_K{}^*)^T$ which is a single compromise solution from the functional efficient solution set F^*.

3. Applied continuous-discrete Optimization Algorithms

Two different ways may be applied to solve continuous-discrete optimization problems.

(a) Obviously, continuous optimization algorithms, which are available in a large number, may be employed. Yet an additional strategy is necessary, to consider several discrete design variables.

(b) On the other hand, there exists several integer optimization algorithms, which may be extended to solve mixed optimization problems as well.

The optimization algorithms described in the following have been applied on structural optimization problems. An important measure of their applicability is the total number of function evaluations (function calls) to achieve an optimal solution, because structural analyses are often time-consuming.

3.1 Continuous Algorithms with Enumeration or Penalty Method

Method 1: If the mixed optimization problem can be treated as if it is purely continuous, a continuous algorithm can be employed on the condition that a structural analysis yields results for the objective functions and the constraints for any one single continuous value of x_{dj}. Finally the solution is rounded according to the discrete design sets of x_d. Unfortunately infeasible solutions are often obtained. According to Fig. 3.1 this may be overcome by performing a partial enumeration within the discrete design set, while keeping the continuous design variables constant.

In doing so, every continuous optimization algorithm can be employed. On the other hand, an additional enumeration is time-consuming, even if it is restricted to a small set of discrete values, because the optimization problem may contain many discrete design variables.

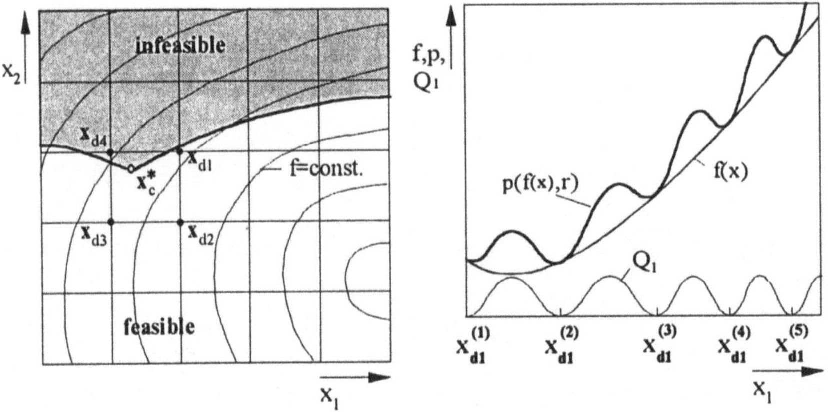

Fig. 3.1. Rounding and partial enumeration

Fig. 3.2. Penalty function and enhanced function of the penalty method

Method 2: The well-known penalty function method for the unconstrained minimization of constrained problems can be utilized to solve discrete or mixed optimization problems [3]. The main intention is to formulate an enhanced function p in place of the objective function (see Fig. 3.2) which will then include the conditions of discrete design sets, such as

$$p = f(x) + r^{(s)} \sum_{j=1}^{N_d} Q_j(x_{dj}) \tag{6}$$

where $r^{(s)}$ penalty parameter of iteration s,

           Q_j penalty function of the discrete variable x_{dj}.

As with method 1, every continuous optimization algorithm can be employed with the penalty function method. But in practice, difficulties will occur with high-grade algorithms, as for example sequential quadratic programming algorithms (SQP). The substitute function (6) causes these algorithms to come to a dead stop at a subminimum of the optimization problem.

3.2 SLP-Algorithm with Partial Enumeration

Method 3: If a sequential linear optimization algorithm (SLP) is applied, the disadvantages of method 1 can be overcome. Here the opportunity is given to carry out an enumeration within the linear subproblem of each iteration. After the solution of the linear subproblem, a partial enumeration within the linearized problem is performed, while the continuous design variables are kept constant. Therefore, each iteration results in a nearly optimal continuous-discrete solution of the linearized subproblem. It is advantageous that no additional structural analyses are necessary, because the linearized objective function and constraints are already known. Additional strategies for finding the best continuous-discrete solution of the linearized subproblem can easily be implemented.

3.3 Integer Gradient Method

Method 4: The integer gradient method [2] is analogous to the steepest descent approach. An important difference, however, is that the search directions are achieved by means of integer gradients. The special rules for the calculation of integer gradients guarantee that the conditions of discrete design values are satisfied, even whilst evaluating the gradients. Amir and Hasegawa [4] show that the approach is applicable to mixed continuous-discrete problems. It is assumed that the continuous design variables can be regarded as if they are discrete variables with discrete subsets X_j. These subsets are part of the continuous design space, but contain only a finite number of values with small increments Δx_{cj}.

First the optimization problem has to be transformed into an unconstrained problem with the substitute objective function p, by means of a penalty function which covers the constraints of the original problem. The algorithm contains different search strategies, see Fig. 3.3, which have the following priority.

(a) Integer gradient search (IG-search): in the direction of the integer gradient $d^{(s)}$, an unidirectional line search is performed to find the optimal step width λ^* of the problem

$$\min_{\lambda \in \mathbb{N}} \left\{ p \left[f(x^{(s)} + \lambda^{(s)} \cdot d^{(s)}) \right] \right\} \tag{7}$$

(b) Subsequential search (SS-Search): this search refers to discrete points in the design space which lie near the integer gradient search direction but are not covered by (7).

(c) Neighbourhood point search (NP-search): in the neighbourhood of a discrete solution, a partial enumeration is performed. In addition, the algorithm may evaluate a new integer gradient direction by using the best neighbourhood point and the solution point of one search before (NPG-search).

Fig 3.3 illustrates the progress of the described method within a two-dimensional design space.

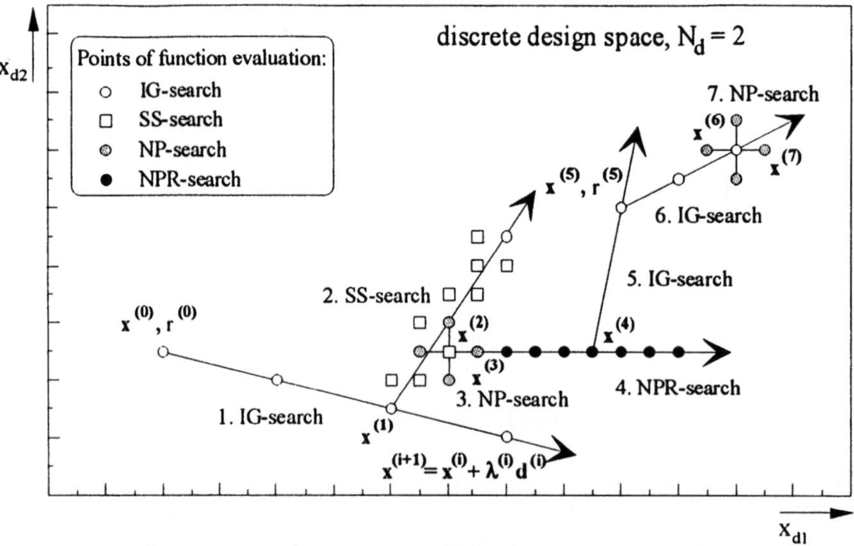

Fig 3.3. Iteration history of the integer gradient method for a two-dimensional example

4. Applications on Composite Structures

4.1 General Optimization Model for Composite Structures

Composite material offers a wide range of design and combination facilities. In this case a general optimization model for composite structures, see [3], is proposed. The following model types may be distinguished.

(a) Material model: as summarized in Table 4.1, the material model covers all design variables which are useful to describe the selection of each material component. In particular, reference is made here to material attributes which are dependent on time and temperature.

(b) Layer model: the layer model deals with the individual construction of a composite layer. Design variables are layer attributes, such as fibre orientation or layer thickness. Additional variables can be found in Table 4.1.

(c) Geometric model: in analogy to the shape optimization of isotropic structures, the geometric model contains those design variables which describe the shape of the composite structure.

For practical applications, design variables of the different models described above will occur simultaneously in one optimization model.

Table 4.1. Exemplary design variables of the material model and the layer model

Type of variable :	continuous	discrete	integer
Material model: Fibre volume ratio	Φ_F	Φ_F	-
Type of fibre	Young`s modulus, E_F = function of ε	-	fibre of carbon, glass, aramid etc.
Type of matrix	Young`s modulus, E_M= function of time,temperature	-	resin of epoxy, polyester etc.
reinforcement	-	-	unidirectional, crossed, symmetrical
Layer model: Layer thickness $t = (t_1, t_2, .., t_N)^T$	(a) linear model $t = A\,x + t_0$ (b) geometric model t = functions of x	t_i out of production tables	constant total thickness with variation of the total number of layers N
Amount of Layers	-	-	N
Fibre orientation $\alpha=(\alpha_1, \alpha_2, ..., \alpha_N)^T$	(a) linear model $\alpha = B\,x + \alpha_0$ (b) geometric model α = functions of x	α_i out of production tables	-

4.2 Composite Plate

A quadratic sandwich plate, simply supported, and under constant pressure p_0 (see Fig 4.1) is to be optimized relative to its weight $f_1=G$ and its maximum displacement $f_2=W$.

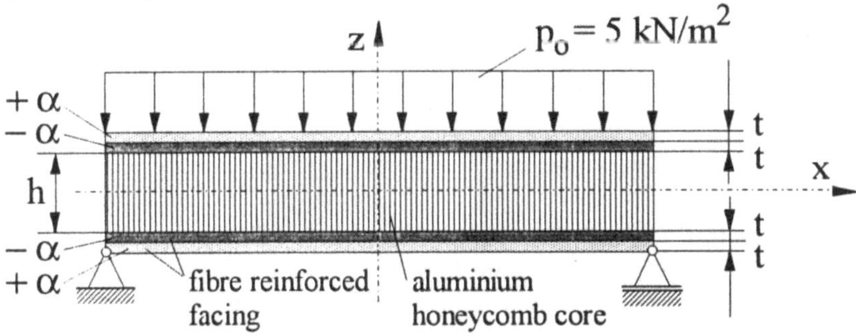

Fig. 4.1. Sandwich plate under constant pressure p_0 (dimension 1000mm x 1000mm) and the discrete subsets $X_1:=\{0°, 5°, 10°, ..., 90°\}$, $X_2:=\{0.25\text{mm},0.3\text{mm},0.35\text{mm}, ...,2.5\text{mm}\}$, $X_3:=\{10\text{mm}, 11\text{mm}, 12\text{mm}, ..., 40\text{mm}\}$

68

The design variables of this vector optimization problem are the fibre orientation α, the layer thickness t and the height h of the aluminium core. Several inequality constraints ensure that no breaking of the fiber or matrix occurs and that the core will not be damaged by shear. The structural analysis is performed by finite element methods.

The vector optimization problem is solved by means of the bounded objective method [4]. A comparison between the continuous functional efficient boundary and some discrete solution points (1-8) is given in Fig 4.2. The discrete solutions come very near to the choosen objective bounds $b^{(i)}$ and the boundary F^*, because the discrete values of the design sets are close together.

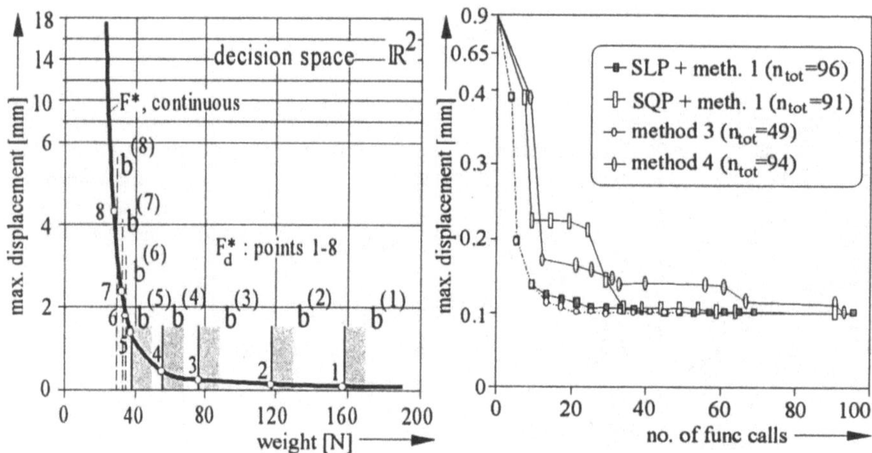

Fig 4.2. Continuous and discrete functional efficient solutions

Fig. 4.3. Iteration history of point 2

An exemplary iteration history for solution point 2 is plotted in Fig. 4.3 for the methods proposed. As it happens, the penalty method (method 2) fails. The comparison between the applied methods shows that the smallest amount of function evaluations, namely 49, is achieved by method 3 (SLP-algorithm with partial enumeration) whereas the purely continuous solution with the same algorithm requires 69 function calls.

4.3 Curved Composite Beam

As presented in [3], a curved composite beam is to be optimized relative to the material costs (objective function f_1=C) and the maximum displacement (objective function f_2=W) which is analytically calculated. Load and displacement conditions are illustrated in Fig 4.4. In addition, Fig 4.4 shows the layer thicknesses of the curved and the straight part of the beam (index c and b) and the fibre orientation, which is identical for both parts. Fibres made of carbon or glass are employed. The model results in the following for continuous and seven discrete design variables of $\mathbf{x} = (\mathbf{x_c}^T, \mathbf{x_d}^T)^T$,

where $\quad \mathbf{x_c} = (r_i, b, \pm_2\alpha, \pm_3\alpha)^T$ and $\mathbf{x_d} = (_2t_b, _2t_c, _6t_c, _3t_b, _3t_c, _5t_c, _4t_b)^T$

and the following details referring to the design space

$$N_c = 4, \ \mathbf{x_l}=(30mm, 120mm, 0°, 0°)^T, \ \mathbf{x_u}=(100mm, 160mm, 90°, 90°)^T,$$
$$N_d = 7, \ X_1 = X_2 = X_3 := \{0.1mm, 0.2mm, 0.3mm, ..., 2mm\},$$
$$X_4 = X_5 = X_6 := \{0.5mm, 1.0mm, 1.5mm, ..., 30mm\},$$
$$X_7 := \{0.4mm, 0.8mm, 1.2mm, ..., 2.4mm\}.$$

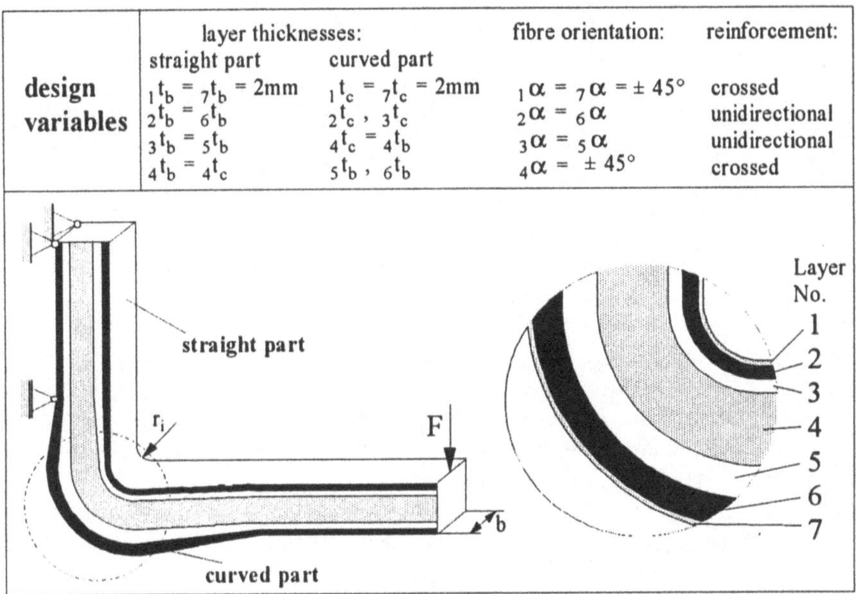

Fig. 4.4. Curved composite beam with the used layer construction

Apart from several couplings of the layer thicknesses, as listed in Fig 4.4, the restrictions for the maximum layer thickness as well as for the total beam thickness $t_{tot}=\Sigma_i t$ are given. Additional inequality constraints consider the break-down criteria of the used fibre and matrix materials, see [3].

The continuous functional efficient solutions are achieved by means of the bounded objective method (trade-off method). The continuous counterpart of the continuous-discrete solution point 4 in Fig. 4.5 is reached after 103 function evaluations, whereas the continuous-discrete solution of method 3 leads to a smaller amount of 90 function calls.

Here, the application of method 1 (SLP algorithm with partial enumeration within the original problem) is very time-consuming, because the total amount will come to no less than $103+3^7=2290$ function evaluations.

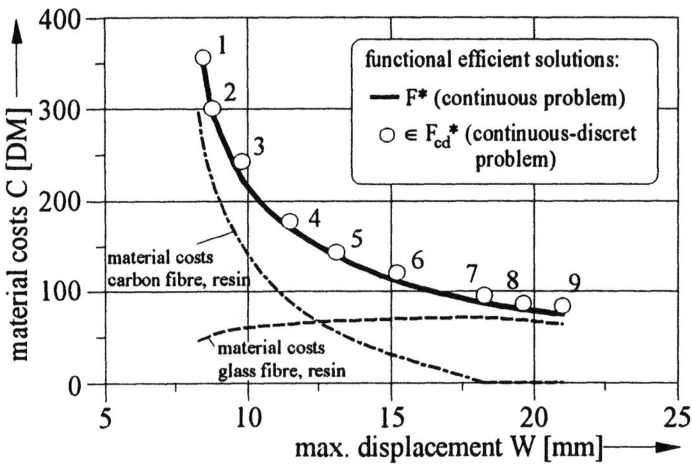

Fig. 4.5. Continuous and continuous-discrete functional efficient solutions

5. Conclusion

If the mathematical representation of the original optimization problem is satisfactory, discrete design variables should be only considered within the subproblem of the optimization algorithm, as is proposed by method 3 or with dual algorithms. The assumption is, however, that the continuous-discrete optimization problem can be lead back to a purely continuous problem.

Otherwise, an algorithm of such a kind should be employed which profits from gradient information or works with a qualified line search direction (see method 4). A total enumeration, as proposed for method 1, is much too time-consuming than to be applied in structural optimization.

Method 2 (penalty method) offers a compromise between method 1 and method 3, but is very sensitive to the continuous optimization algorithm employed and the given problem.

References

[1]   Eschenauer,H.; Geilen, J.: Strukturanalyse und Optimierung dickwandiger, gekrümmter Faserverbundbauteile, ZAMM 71, 6T713-T715, 1991

[2]   Eschenauer, H.; Schäfer, E.: Über die Lösung von Strukturoptimierungsaufgaben mit Diskretheitsforderungen, ZAMM 70, 4T279-T281, 1990

[3]   Geilen, J.: Optimierungsmodellbildung und Auslegung dickwandiger, gemischter Composite-Bauweisen mit stark gekrümmten Bereichen, Dissertation, Universität-GH Siegen, 1992

[4]   Schäfer, E.: Interaktive Strategien zur Bauteiloptimierung bei mehrfacher Zielsetzung und Diskretheitsforderungen, Dissertation, Universität-GH Siegen, Fortschr.-Ber. VDI Reihe 1, Nr. 197, VDI Verlag, Düsseldorf 1990

A Random-Search Approach to Multicriterion Discrete Optimization

Andrzej Osyczka[1] and Jerzy Montusiewicz[2]

[1] Department of Mechanical Engineering, Technical University of Cracow, 31-864 Cracow, Jana Pawła II Ave., 37, Poland

[2] Department of Technology Foundation, Technical University of Lublin, 20-618 Lublin, Nadbystrzycka Str.,36, Poland

Abstract. In the paper the random search method is used for generating a set of feasible solutions for general nonlinear programming models with discrete decision variables. From this set all Pareto optimal solutions are selected using the algorithm based on contact theorem. For most engineering design problems the set of Pareto optimal solutions obtained in this way is large and it is difficult and tiresome to make the right decision while looking through all of them. Thus a method for selecting a preferred subset of Pareto optimal solutions is proposed. The computer program for this method has been developed. The program has an interactive character and all optional parts of the method are run on the basis of a dialogue with the computer.   Finally the real - life application example dealing with an optimum design of a gear set is discussed.

Keywords. Multicriteria optimization, random search,Pareto optimal set, discrete model, gear design.

1 Introduction

Solving nonlinear single criterion optimization models with discrete decision variables is a fairly difficult task [1]. The problem becomes more complicated when multicriterion optimization models are considered [2],[3]. In this case the solution of the problem is represented by a set of Pareto optimal solutions (nondominated solutions) and the problem is to find all of them or some representative subset. If the set of Pareto optimal solutions is large [4]. In the paper a method for seeking a preferred subset of Pareto optimal solutions for discrete nonlinear models is proposed.

The problem under consideration is formulated as follows: find $x = [x_1, x_2,..., x_n]^T$ such that

$$f(x^*) = \min_{x \in R^n} f(x) \tag{1}$$

and such that

$$g(x) \geq 0 \tag{2}$$

72

under assumption that $x_i \in X_i$ for i=1,2,...,n,

where: $\mathbf{x}^* = [x_1^*, x_2^*,..., x_n^*]^T$ is the vector of decision variables,

$\mathbf{f(x)} = [f_1(x), f_2(x),..., f_k(x)]^T$ is the vector of objective functions,

$\mathbf{g(x)} = [g_1(x), g_2 x),..., g_m(x)]^T$ is the vector of inequality constraints,

X_i - is the given set of discrete values of the i-th decision variable.

2 Method of solution

The method consists in two stages. At the first stage the set of all Pareto optimal solutions is generated using the random search method. The aim of the second stage is to select the preferred subset of Pareto optimal solutions on the basis of which it is easy to make the final decision as to the choice of the final solution with the respect to designer's preferences.

2.1 Method for generating the set of Pareto optimal solutions

Using the random search methods the set of Pareto optimal solutions can be generated as follows:
1. Generate a random point $\mathbf{x}^1$ using the formula:

$$x_i^1 = x_i^1 + r_i (x_i^u - x_i^1) \qquad \text{for i=1, 2,..., n} \qquad (3)$$

where: x_i^1 - lower bound on the i-th decision variable,

x_i^u - upper bound on the i-th decision variable,

r_i - random number between zero and one.

2. Round the value of x_i^1 to the nearest value from the discrete set X_i , for i=1, 2,..., n.
3. If the point $\mathbf{x}^1$ satisfies constraints (2) go to 4. Otherwise go to 1.
4. Evaluate the values of the objective functions for the point $\mathbf{x}^1$, i.e., evaluate $f_1(\mathbf{x}^1), f_2(\mathbf{x}^1),..., f_k(\mathbf{x}^1)$.
5. If the point $\mathbf{x}^1$ represents a Pareto optimal solution add it to the existing set of Pareto optimal solutions and go to 1. Otherwise go straight to 1.

The procedure is repeated until the assumed number of points L is generated. For realization of point 5 of this method PARETO algorithm as presented in [4] is developed. As the result of the above procedure T Pareto optimal solutions are obtained. From the theoretical point of view all these solutions are nondominated and the choice of one of them belongs to the decision maker. This choice may be difficult and tiresome if T represents a large number, which is a typical situation in many engineering tasks. Thus the method of selecting a representative subset of Pareto optimal solutions is presented below.

2.2 Method for selecting the preferred subset of Pareto optimal solutions

The method consists of three independent steps which can be used optionally depending on the knowledge the decision maker has at his disposal. In the extreme case, he may only state the number of solutions he wants to be output and the method will select them in the way that they will be more or less evenly distributed on the plane designated by the whole set of Pareto optimal solutions.

Step 1 - Displacement of the ideal vector.
Each multicriterion optimization problem is characterized by an ideal vector which represents the separately attainable minima and which will be denoted by $f^\circ = [f_1^\circ, f_2^\circ, ..., f_k^\circ]^T$, where:

$$f^\circ_i = \min_{x \in X} f_i(x) \qquad (4)$$

where: X is the set of feasible solutions.
In this step the decision maker may want to restrict the set of Pareto optimal solutions to the boundary defined by the given vector $f^{o\prime} = [f_1^{o\prime}, f_2^{o\prime}, ..., f_k^{o\prime}]^T$ for which for every i, $f_i^{o\prime} \leq f_i$. The method will discard all the solutions which are out of the boundary defined by the new ideal vector $f^{o\prime}$ (see Fig.2.1).

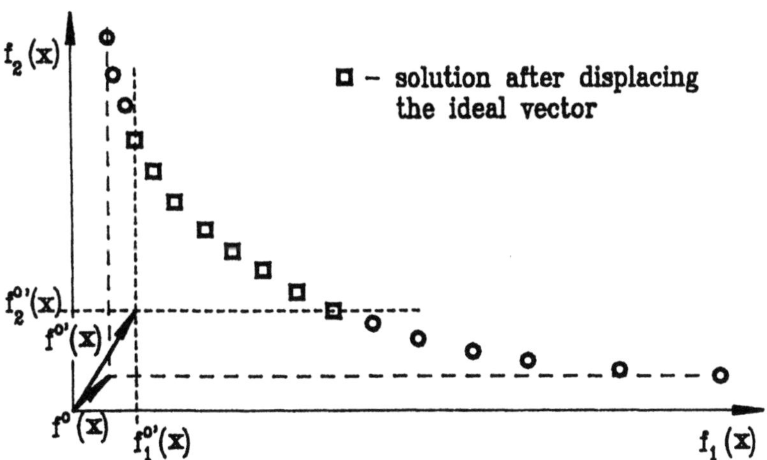

Fig.2.1 Example of the displacement of the ideal vector for bicriterion minimization

<u>Step 2</u> - Defining the undifferentiation threshold.
The designer fixes the value of the undifferentiation threshold u in percentage for all the criteria,for example 1.5 % or 2 %. If there a more solutions than one within the assumed theshold u only one of them is selected (see Fig.2.2).

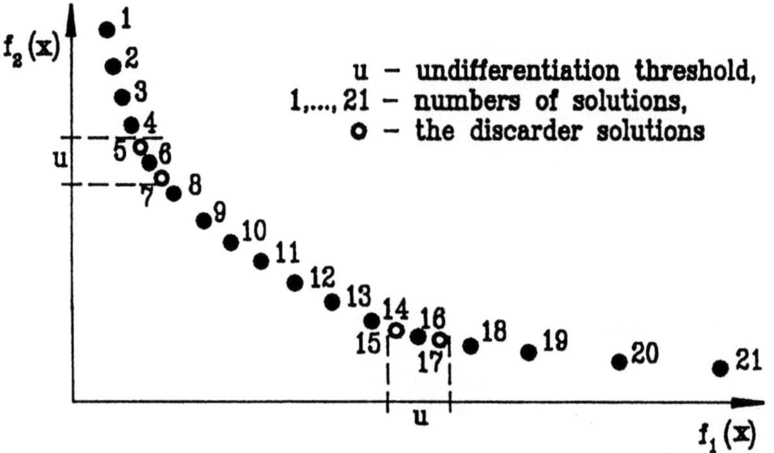

Fig.2.2 Example of applying the undifferentiation threshold for bicriterion minimization

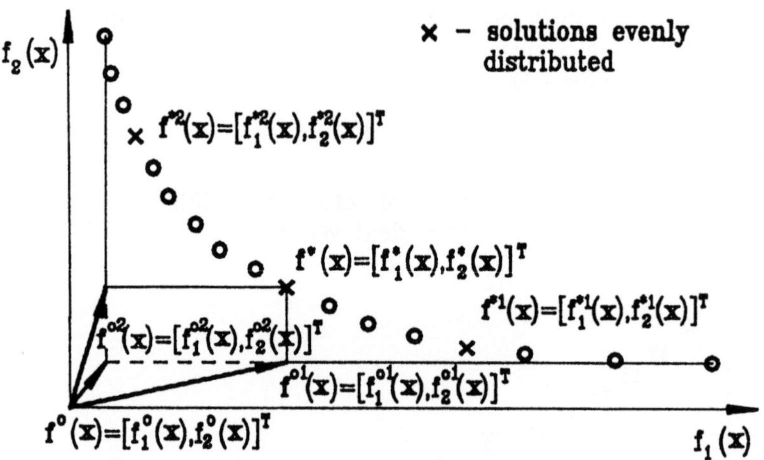

Fig.2.3 Calculation of representative solutions using the min-max approach for bicriterion minimization

<u>Step 3</u> - Searching a representative subset of solutions.
In this step the min-max approach is used for finding a representative subset of Pareto optimal solutions. The method consists of several phases depending on the number of solutions the designer wants to select. In phase 1 the solution optimal in the min-max sense is found.

For this solution the vector of the objective functions is $\mathbf{f}^*$. In phase 2 the k-th ideal vectors are created using the vectors $\mathbf{f}^\circ$ and $\mathbf{f}^*$. For each of these vectors the solution optimal in the min-max sense are found. In this way k-solutions are selected. The procedure is repeated and in the n-th phase n·k ideal vectors are created and n·k solutions are selected and added to the existing set.

The idea of this approach is shown in Fig.2.3 where a bicriterion optimization problem is considered.

The general diagram of the method described above is presented in Fig.2.4. On the basis of this diagram the computer program has been developed. The program is coded in FORTRAN and can be treated as a part of Computer Aided Multicriterion Optimization System (CAMOS) [6]. The structure of the program is similar to the structure of CAMOS, i.e., the program is run on the basis of the dialogue with the computer.

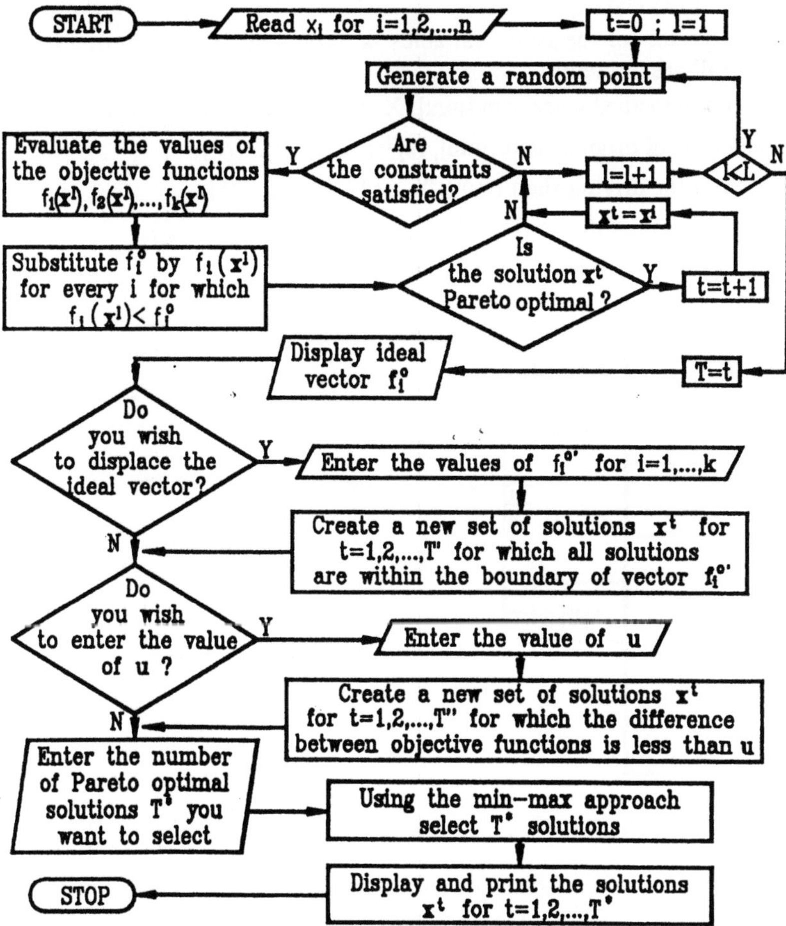

Fig.2.4 General flaw diagram of the method

3 Application to gear set design

3.1 Optimization model

The method of generating a Pareto optimal set and the method for selecting a representative subset of solutions is applied to solving the problem of designing a gear set. Due to page limitation only a short description of the optimization model is provided here. This model is based on formulae presented in the book [5] which are similar to those, presented in [7]. The scheme of the gear set under consideretion is presented in Fig.3.1.

Elements of a vector function $\mathbf{f(x)}$ represent the following objective functions:

$f_1 (\mathbf{x})$ - volume of the gear [mm^3],

$f_2 (\mathbf{x})$ - distance between axes [mm],

$f_3 (\mathbf{x})$ - width of toothed rim [mm].

The vector of decision variables $\mathbf{x}$ represents the following overall dimensions (all are discrete):

x_1 - width of the toothed wheel rim [mm], $X_1 = \{24, 24.5, 25, 25.5, ..., 42\}$,

x_2 - diameter the of driving shaft [mm], $X_2 = \{32, 33, 34, ..., 70\}$,

x_3 - diameter of the driving shaft [mm], $X_3 = \{48, 49, 50, ..., 95\}$,

x_4 - number of teeth of the driving wheel [-], $X_4 = \{14, 15, 16, ..., 25\}$,

x_5 - module pitch [mm], $X_5 = \{4, 4.5, 5, 5.5, 6\}$.

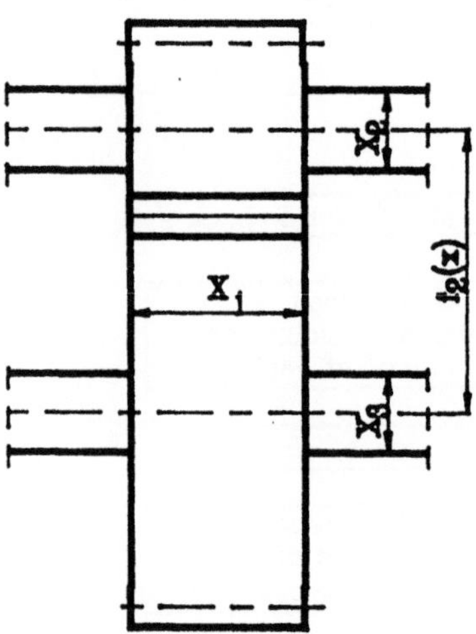

Fig.3.1 Scheme of the gear set

The decision variables are restricted by 13 inequality constraints which refer to:

- geometrical dimensions of the gear,
- permissible stress in the shaft and in the intermeshing teeth.
       The basic data of the gear are:
- basic data: input power, 12 [kW], rotation input speed, 280 [rev./min.], transmission ratio, 0,317, allowable deviation of the transmission ratio, 0.01,
- material data: allowable bending stress of the driving and the driven wheel, 105 [MPa], allowable surface pressure for the driving and the driven wheel, 62 [MPa], allowable torsional strength of the shaft, 70 [MPa],

3.2 Optimization results

<u>Step 1</u> - First the set of Pareto optimal solutions is generated, it has the 22 elements, the ideal vector $f^\circ = [307140 \ 145 \ 24]^T$. To restrict the set of Pareto optimal solutions a new ideal vector $f^{o*} = [310000 \ 145 \ 25]^T$ is assumed. For this ideal vector the Pareto optimal set is reduced to 13 solutions.

<u>Step 2</u> - Assuming the value of the undifferentiation threshold for all criteria $u = 1,7\%$. The Pareto optimal set is further reduced to the 11 elements.

<u>Step 3</u> - For finding a representative subset of Pareto optimal solutions the min--max approach is used. In this way four following solutions are obtained:

Solution **a**

Vector of objective functions
$$f^*(x) = [333536 \ 174,00 \ 29,5]^T$$
Vector of decision variables
$$x_1 = 29,5, \ x_2 = 38, \ x_3 = 54, \ x_4 = 21, \ x_5 = 4.0$$

Solution **b**

Vector of objective functions
$$f^{*1}(x) = [340522 \ 182,00 \ 27,0]^T$$
Vector of decision variables
$$x_1 = 27,0, \ x_2 = 39, \ x_3 = 54, \ x_4 = 22, \ x_5 = 4.0$$

Solution **c**

Vector of objective functions
$$f^{*2}(x) = [313421 \ 186,75 \ 26,5]^T$$
Vector of decision variables
$$x_1 = 26,5, \ x_2 = 37, \ x_3 = 53, \ x_4 = 20, \ x_5 = 4.5$$

78

Solution **d**

Vector of objective functions
$$\mathbf{f}^{*3}(\mathbf{x}) = [347534 \quad 166,00 \quad 32,5]^{\mathrm{T}}$$

Vector of decision variables
$$x_1 = 32,5 \,, \quad x_2 = 38 \,, \quad x_3 = 55 \,, \quad x_4 = 20 \,, \quad x_5 = 4.0$$

The graphical representation of these solutions is shown in Fig. 3.2. From this set of solutions the designer can easlly make the right decision.

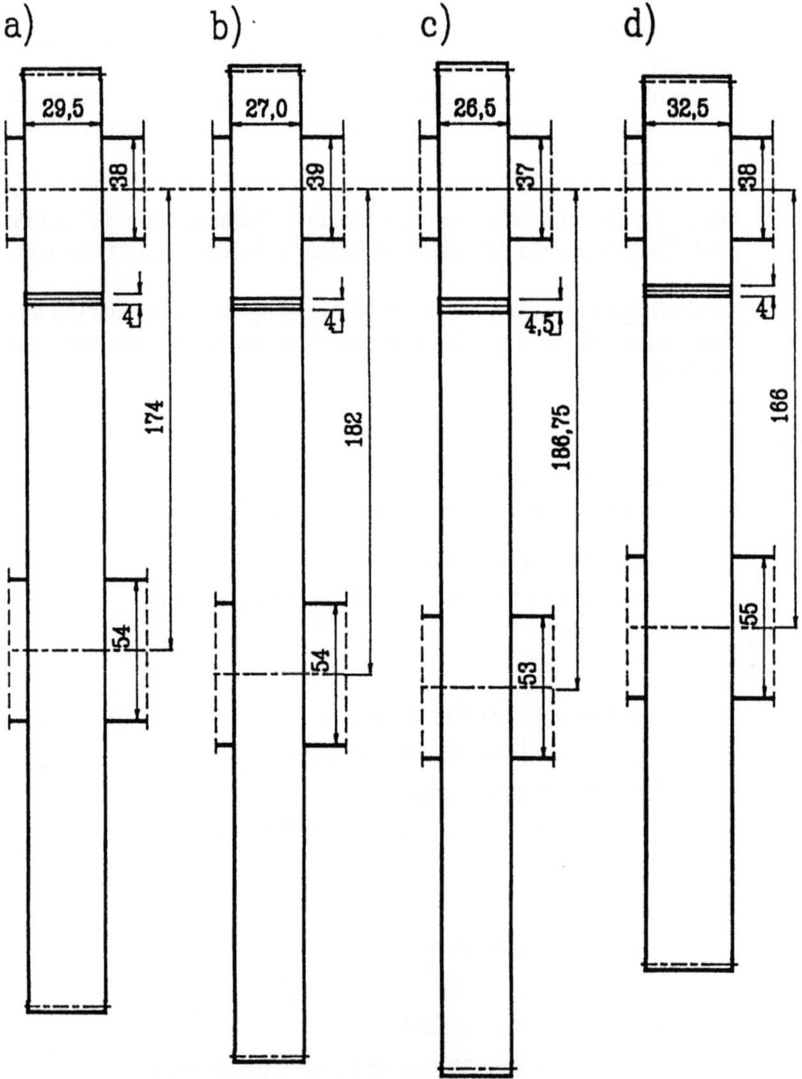

Fig.3.2 Graphic outline of the optimal solutions

4 Conclusions

The method discussed solves multicriterion optimization problems for nonlinear models with discrete decision variables providing the designer with a preferred subset of Pareto optimal solutions, which are helpful in making the right decision in conflicting situations.

The method does not require the information about the preferences as to importance of the criteria. The obtained subset of solutions is generated almost automatically and the process of generation reqiures only the information the designer is able to provide.

Although the method is designed to solve the discrete models there is no restriction in using it for the other models, i.e., the models with continuous and mixed continuous-discrete variables. The computer program is prepared to solve all these models.

The only disadvantage of this method is that the number of decision variables considered in the model can not be large. This is the general restriction for all random search methods.

References

1. Paczkowski W.M.: OPTIM - Nonlinear Discrete Optimization Program, 10th Conference on Computer Methods in Mechanics, Świnoujście 1991, Poland
2. Hajela P.,Shih C.J.: Multiobjective Design in Mixed Integer and Discrete Design Variable Problems, Proceeding AIAA/ASME/AHS/ASC 30th Struct. Dynamics and Materials Conference, Mibile, Alabama, April 3-5, 1989
3. Sobieski-Sobieszański J.: Optimization by Decomposition in Structural and Multidisciplinary Applications, In: Optimization of Large Structural Systems NATO/DFG ASI Conference Berctesgaden, 1991, Germany
4. Rosenman M.A.,Gero J.S.: Reducing the Pareto Optimal Set in Multicriteria Optimization, Eng. Opt., vol. 8, 1985, pp. 189-206
5. Osyczka A.: Multicriterion Optimization in Engineering with FORTRAN Programs, John Wiley and Sons, Chichester, 1984
6. Osyczka A.: Computer Aided Multicriterion Optimization System (CAMOS) -Software Package in FORTRAN, International Software Publishers, Cracow, 1992
7. Thwberge Y.,Cardou A., Cloutier L.: Parallel Axes Gear Set Optimization in Two-Parameter Space, MPT'91 JSME Int. Conf. on Motion and Power Trans., Nov. 23-26, 1991, Hiroshima

Optimal Arrangement of Ribs in Rib-Reinforced Plates and Shells

A. O Rasskazov, Dr., Prof.[1] and A. S. Dekhtjar, Dr., Prof.[2]

[1] Department of Theoretical and Applied Mechanics,
Kiev Institute of Automobiles and Highways
Ul. (*street*) Suvorova 1, Kiev, 262010, UKRAINE

[2] Department of Structural Mechanics
Ukrainian Academy of Art, Architectural Department
Ul. (*street*) Smirnova - Lastochkina 20, Kiev, 252053, UKRAINE

1. Introduction

Rib reinforcement of plates and shells is commonly used in machinery, in building structures and in other branches of technology. A problem of an optimal arrangement of ribs in thin rigid-plastic plates with hinged edges was studied in the paper[1] in such a formulation: a volume V of material in an initially non-reinforced structure of constant thickness is presented, it is necessary so to distribute an added volume αV as to obtain the construction of greatest load carrying capacity.

Constructions of variable thickness will be excluded here as unsuitable for technological reasons. Thus two possibilities arise. The first of them may be a plate or a shell of constant (augmentable) thickness. The other may involve stiffening by ribs. In a general case we can devise quite a few patterns of rib arrangement and it is rarely possible to transform one pattern into another by some continuous change of one or several parameters. Therefore such patterns may be considered as independent data *a priori* given in a design optimization problem.

In addition, in part 2 of the present paper we consider a square plate supported at its four corner points only, the external load being uniformly distributed. In part 3 a shallow shell of revolution with radial, circular or spiral ribs is considered.

2. Plate

A square plate of constant thickness h has the side length 2b. Its material is rigid - perfectly plastic and in general with differing yield limits σ_o in tension and $\rho\sigma_o$ - in compression. The intensity of the distributed load is denoted by q. We also introduce the notations: $p = q\sigma_o^{-1}$; $h = \varepsilon b$.

When $\rho = 1$ the upper bound of dimensionless limit load intensity may be expressed as

$$p = 0.5\varepsilon^2 \tag{1}$$

and the volume of the material is given by $V = 4b^3c$.

Our first case is that of a plate with constant thickness so the upper bound of the dimensionless limit load becomes

$$p = 0,5\varepsilon^2(1+\alpha)^2 \tag{2}$$

In order to compare results obtained below with the same results in paper[1] we express them in the form

$$p = 0,5\varepsilon^2\phi(\alpha,\varepsilon)^2 \tag{3}$$

Thus for the plate of constant thickness (2.2) we have:

$$\phi_o(\alpha,\varepsilon) = (1+\alpha)^2/3 \tag{4}$$

while for the unreinforced plate it is:

$$\phi(\alpha,\varepsilon) = 1/3$$

2.1 The First Rib Arrangement Pattern

We assume that all the ribs have a rectangular cross section with a height h_r and a width b_r. The ribs are symmetrically situated with respect to the middle surface of the plate. Let $b_r = \beta h_r$ and $h_r = \nu b$, so $b_r = \beta \cap b$.

Four ribs are arranged along each side in the first pattern. A volume of the added material contained in the ribs is equal to $8b^3\beta\nu^2$. Taking into account the expression $V_r = 4b^3\varepsilon\alpha$ it follows that

$$\nu = (\alpha\beta^{-1}\varepsilon/2)^{1/2} \tag{5}$$

2.1.1 The First Collapse Mode

Boundary conditions and a load distribution allow to consider several failure modes. The first of them is shown in fig. 2.1,a.

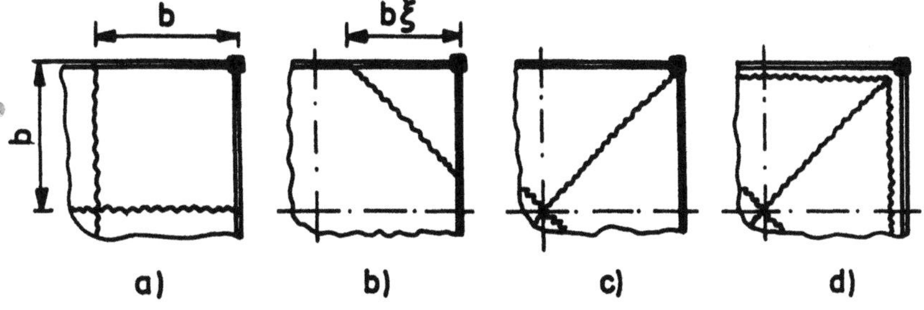

Fig. 2.1

82

The kinematical method of limit analysis is used in the form of the yield line theory so we can calculate the energy dissipation D_1 and external force power D_e.

Taking into account the notations introduced above one can obtain

$$D_1 = \sigma_o b^3(\varepsilon + \nu^2\beta); \quad D_e = 2p_{11}\sigma_o b^2,$$

then

$$p_{11} = 0.5(\varepsilon^2 + \nu^3\beta) \tag{6}$$

According to (1) and (3) we can finally see from (6)

$$\phi_{11}(\alpha,\beta,\varepsilon) = 1/3 + \alpha(\alpha\beta^{-1}\varepsilon/2)^{1/2}/6\varepsilon. \tag{7}$$

2.1.2 The Second Failure Mode

This form is shown in fig. 2.1,b. It has an indefinite value $b\xi$ in a corner portion of the plate. Therefore the dissipation D_1 and the external load power D_e are functions of the variable ξ

$$D_1 = \sigma_o b^2 \frac{\xi\varepsilon^2 + 2\nu^3\beta}{2\xi}; \quad D_e = p\sigma_o b^2 \frac{6-\xi}{6}.$$

One could obtain the dimensionless limit load intensity from these expressions

$$p_{12} = \frac{3(\xi\varepsilon^2 + 2\beta\nu^3)}{\xi(6-\xi^2)} \tag{8}$$

and in accordance with the kinematical method of limit analysis it is necessary to minimize p_{11} by ξ. A value ξ which minimized the expression (8) depends on the parameters β, ε and ν. For instance when $\beta = 0.5$, $\nu = 0.1$ and $\varepsilon = 0.05$ one finds $p_{12\,min} = 2.038$ and $\xi = 0.8$ but if $\nu = 0.2$ then $p_{12\,min} = 2.1$ and $\xi = 1$.

By taking (3) and (5) into account the following expression can be obtained for the second failure mode

$$\phi_{12}(\alpha,\beta,\varepsilon) = \min_{\xi} \frac{2[\xi + \alpha(\alpha\beta^{-1}\varepsilon/2)^{1/2}/\varepsilon]}{\xi(6-\xi^2)} \tag{9}$$

2.1.3 The Third and the Fourth Failure Modes

These modes are shown in fig. 2.1,c and d. The first of then means failure of ribs and plate, the other is only concerned with plate failure. These modes have been studied in [1] so we can write the final expressions

$$\phi_{13}(\alpha,\beta,\varepsilon) = 8[2 - (\alpha\beta^{-1}\varepsilon/2)^{1/2}]^{-1};$$

$$\phi_{14}(\alpha,\beta,\varepsilon) = 1 + (0,125\alpha^3\beta^{-1}\varepsilon^{-1})^{-1}. \tag{10}$$

In order to obtain the final estimate of the limit load of a stiffened plate it is necessary to compare the right-hand parts of the expressions (4), (7), (9) and (10).

A preliminary analysis of these expressions allows us to exclude the estimate ϕ_{14} by virtue of $\phi_{14} = 3\phi_{11}$. One can finally write for the first rib pattern

$$p_1 = 1,5\varepsilon^2\phi_1; \quad \phi_1 = \min(\phi_{11}, \min_\xi \phi_{12}, \phi_{13}). \tag{11}$$

2.2 The Second Rib Pattern Arrangement

Now we consider a pair of diagonal ribs with the same initial assumptions as in part 2.1. It may be shown that

$$\nu = (2^{-1/2}\alpha\beta^{-1}\varepsilon)^{1/2}. \tag{12}$$

2.2.1 The First and the Second Failure Modes

In these cases failure modes coincide with the same modes in parts 2.1.1 and 2.1.2. Therefore we can write

$$\phi_{21}(\alpha, \beta, \varepsilon) = 1/3 + \frac{\alpha}{6\epsilon}(2^{-1/2}\alpha\beta^{-1}\varepsilon)^{1/2} \tag{13}$$

$$\phi_{22}(\alpha, \beta, \varepsilon) = \min_\xi \frac{2\xi + \alpha(2^{-1/2}\alpha\beta^{-1}\varepsilon)^{1/2}}{\xi(6 - \xi^2)} \tag{14}$$

The estimation ϕ_{22} is similar to ϕ_{12} and should be minimized by ξ. For example if $\beta = 0,5$, $\varepsilon = 0,5$, and $\nu = 0,1$ one can find $\phi_{22\,\min} = 1.092$ and $\xi = 0.7$.

2.2.2 The Third Failure Mode and the Final Estimate

In addition to the two failure modes considered above there may exist a failure mode shown in fig. 2.2. The circular hinge yield line has a radius

$$r = 0,5b(2^{1/2} - \beta\nu). \tag{15}$$

Using a well known estimate for limit load $q = bm_o r^2$

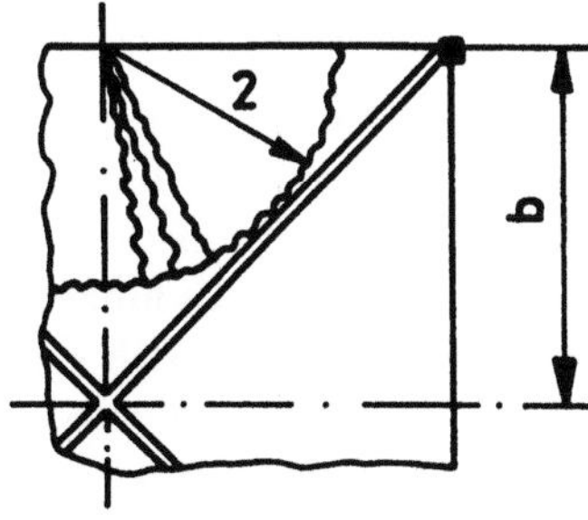

Fig. 2.2

and taking (15) into account, it may be obtained after simple transformation

$$\phi_{23}(\alpha,\beta,\varepsilon) = 4[2^{1/2} - (2^{-1/2}\alpha\beta^{-1}\varepsilon)^{1/2}]^{-2} \tag{16}$$

finally the estimate of load carrying capacity for the plate with diagonal ribs may be written

$$\phi_2 = \min(\phi_{21}, \min_{\xi} \phi_{22}, \phi_{23}). \tag{17}$$

2.3 Optimal Rib Arrangement and Discussion

The next expression represents a solution of an design optimization problem for a ribbed plate

$$\phi_{opt} = \max(\phi_o, \phi_1, \phi_2). \tag{18}$$

Calculations were made according to Eq. (18) for two rib configurations $\beta = 0,33$ and $\beta = 0,5$. Parameters α and ε were chosen from the range $0.1 \leq \alpha \leq 1.5$ and $0.01 \leq \varepsilon \leq 0.05$ in these calculations. Some of the results obtained are represented in table 2.1. An upper entry means ϕ_{opt} and the lower entry of a pair means the optimal pattern of stiffening in each cell of table 2.1 (zero denotes unribbed plate).

Table 2.1

ε	α 0.1	0.2	0.3	0.5	1.0
0.01	0.410	0.550	0.732	1.192	2.145
	2	2	2	2	2
0.02	0.403	0.487	0. 615	0.940	2.050
	0	2	2	2	2
0.03	0.403	0.480	0.564	0.829	1.735
	0	0	2	2	2
0.04	0.403	0.480	0.563	0.762	1.547
	0	0	0	2	2
0.05	0.403	0.480	0.563	0.750	1.419
	0	0	0	0	2

We notice that for the right-hand upper part of the table the optimal pattern of ribs is a diagonal arrangement but also a range of parameters α, ε exists when unreinforced plate of constant thickness is optimal. This conclusion is similar to that for plates with hinged edges obtained before[1].

One can compare the data in table 2.1 with the same data in paper[1] and one can find therefrom that optimally reinforced plates with hinged edges have a load carrying capacity which exceeds by a factor of 2.9 to 4.1 the load carrying capacity of a plate with four point supports. This ratio is exactly 3 for unreinforced plates.

3. Shells of Revolution

A shallow shell of revolution with a constant thickness is considered. Its surface is formed by rotating a t-degree parabola around its axis. The circular edge may be hinged or clamped fR^{-1}, $e = hf^{-1}$. Taking into consideration a shallow form of the shell a transversal load may be distributed arbitrarily along the meridian while preserving axial symmetry.

Notations are introduced in accordance to[2]. R is a radius of a circular edge, f is the largest altitude, $\gamma = a$ shell we can assume its volume of material to be $V = \pi R^3 e\gamma$ and the added volume of material may be expressed as

$$V_r = \pi R^3 e\gamma\alpha \tag{19}$$

A mesh of circular and radial lines is used for discretization. This mesh has n_1 steps along the radial line and $4n_2$ steps in the circumferential direction. We consider three patterns of rib arrangement: a radial, a circular and a spiral one.

The spacings of the radial ribs (fig.3.1,a) in the circumferential direction are denoted by l, so that the number of ribs will be $m = 4n_2 l^{-1}$. Their total length is $L = mR$ and the area of the cross section equals to $F_r = \pi R^2 \alpha e\gamma m^{-1}$.

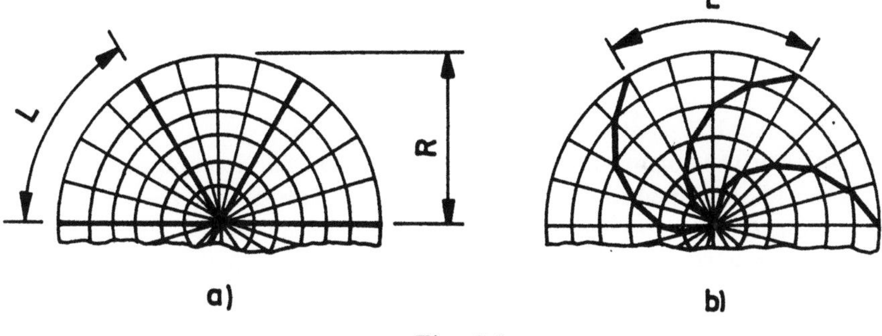

Fig. 3.1

In accordance with the selected mesh the reduced dimensionless height of a rib may be represented as

$$e_r = 2\alpha n_1 n_2 e(mi)^{-1}; \tag{20}$$

In this expression $e_r = h_r(\gamma R)^{-1}$. From (20) it follows that e_r is a function of i which numbers the circular lines of the mesh. For spiral ribs (fig. 3.1,b) and for circular ribs one can similarly find

$$e_r = 2\alpha n_1 n_2 e(2.1mi)^{-1}; \tag{21}$$

$$e_r = e\alpha n_1^2 \left(2l \sum_{k=1}^{n/l} k\right)^{-1} \tag{22}$$

When a system of ribs is formed, the dimensionless value of e_r (in a definite point of a the mesh) must be chosen in accordance with the expressions (20), (21) or (22), it depends upon the ribs pattern.

3.1 Calculations of the Load Carrying Capacity

We make use of the functional obtained in the paper[2] which describes the upper bound of the limit load. This functional is defined over a set of a great number of kinematically admissible fields of displacement velocities. It is based on the generalized Iohansen yield criterion. Further the variational problem is replaced by a problem for minimization of some function of many variables prior to introducing of the discretization mesh.

The upper bound of the total limit load may be expressed as

$$Q = \pi \sigma_o R^2 \min_{w(r,\Theta)} K. \tag{23}$$

When the external load and boundary conditions are axisymmetric one can assume that at any node of the mesh only a normal displacement may exist. These displacements are taken as independent variables.

For each rib pattern arrangement two alternative cases are considered: i.e. total plastic failure of the shell together with some ribs and on the other hand a failure of the shell between ribs only. In this last case we use logical R-functions in order to describe the kinematically admissible fields of displacement velocities. A method for minimization of K (23) was described in[3].

3.2 Numerical Results and Discussion

Results of the calculations are presented for parabolic cupolas i.e. shells with a meridian shaped as a parabola of the second degree. In these examples a rigid - plastic material was assumed to have differing yield limits in tension and in compression (similar in this respect to reinforced concrete). A constant thickness of the shell, its convexity parameter and the magnitude of the added material relative volume α were varied in a wide range. For instance $0,5 \leq \alpha \leq 4,0$.

For the shell with parameters $e = 0,5$, $\gamma = 0,1$ and $\alpha = 0,5$ upper bounds of the limit load $10^3 K$ are presented in table 3.1 (letter "L" means a local failure)

Table 3.1

| The Ribs | The Spacing of the Ribs | | | | | Unreinfor- |
	2	3	4	6	8	ced shell
Radial	0.394	0.394	0.388-L	0.336-L	0.322-L	0.301
Spiral	0.338	0.340	0.342	0.338-L	0.323-L	0.301
Circular	0.321	0.310-L	-	0.247-L	-	0.301

Results similar to presented in tab. 3.1 were obtained for other combinations of parameters and for shells with the equal yield points of material ($\rho = 1$).

The main results obtained are as follows.

1. The lowest effectiveness was found when circular ribs were used only.

2. The system of radial ribs has the greatest effectiveness but only in the case when a total failure of the whole shell together with ribs occurs i.e. when angular spacing of ribs and α-value of the added material are both small. When local failure occurs both radial and circular ribs have a similar effectiveness.

3. If a shell material has equal yield limits both in tension and in compression and value α is fixed, the load carrying capacity of the shell is almost independent of the number of its ribs.

4. If the added material parameter α is relatively small an optimal pattern of radial ribs can be found.

5. When α increases there might apper a tendency excessively to decrease the angular spacing of radial ribs which means that a "smooth" shell with a constant thickness is preferable in this case.

References

1. A. S. Dekhtjar, On the load carrying capacity of rib reinforced stiff - plastic plates. (in Russian) Ukrainian Applied Mechanics (II.M.), (1993) **24(39)**, no. 4, pp. 70-74.

2. A. S. Dekhtjar, Load carrying capacity of axisymmetric shells. (in Russian) Ukrainian Applied Mechanics, (II.M.), (1981), **28**, no 3, pp. 57-60.

3. A. S. Dekhtjar, A. O. Rasskazov, Load carrying capacity of thin-walled structures (in Russian), Kiev, Budivelnik publ. (1990), pp. 152.

Application of the Expert System for Discrete Optimization of Space Truss

Janusz Niczyj[1] and Witold M. Paczkowski[1]

[1]Department of Civil Engineering, Technical University of Szczecin, 70-311 Szczecin, Al. Piastów 50, Poland

Abstract. The paper presents the problems associated with the discrete optimization of halls covered with a space truss. A particular attention is paid to functions which, in the optimization analysis, are performed by the expert system and to tools this system makes use of. The expert module allows the interactive cooperation with optimization programs, facilitates introducting a task, allows the adoption of optimization algorithm control, creates and uses the knowledge base while computing and also handles results. The prepared procedures for the information selection and reduction as well as for the inference and decision - making allow the system knowledge base to "self-teach" and be updated.

Keywords. Discrete optimization, expert system, space truss.

1. Introduction

The paper presents an expert system of controlling a process of optimization. Expert system helps to choose of a preferred vector of control in the orho-diagonal method. The system bases on a dialog with the user [8]. Special attention is payed to time analysis and a change of vector of control during the process of optimization. The vector of control parameters consists of several elements, which provide the information on the feasible region, objective function character, problem starting point, time for computing, optimization algorithm parameters, conditions of stop-page, type of a task performed.

The OPTIM program [8] used for the optimization was developed basing on the algorithm of the O-D (ortho-diagonal) method for seeking minimum of nonlinear function within the discrete feasible domain. The discrete values used in a problem being analysed, are a vector of decision variables of the problem. The task objective function is determined in multidimensional space. Two steps are distinguished in the O-D method: test step and working step. During the test step, the neighbourhood of given point x_i (especially that of starting point x_0) is investigated in order to determine the direction of objective function value correction. Every succesive step utilizes results from the preceding one, checking thereby the neighbourhood more accurately, but yet is more time-consuming. During the working step, minimization of the objective function along the direction of correction takes place in one of three ways possible. The vector x_{ri} that minimizes the value of objective function in the correction direction becomes the starting point x_{i+1} in succesive test step which determines a new direction of correction. This process proceeds up to the moment when, in the neighbourhood of point x_i, the value of objective function cannot be corrected.

2. Expert System of Structure Optimization

In the presented expert system, a knowledge base that stores data on the optimization process was formed on information data which were received from an expert and which resulted from his experience and knowledge [2,4,8,9].

The following are the features of the applied system:
- knowledge base on algorithm controlling made of the information obtained from an expert,
- knowledge acquired by means of decision tables and trees [2,3],
- representation of the knowledge is a based on the principle of base-rules,
- knowledge base updating by the means intermediate solutions of problems being analised as well as selection and optimization of obtained control states,
- dialog with the user are applied, is performed by the means explain procedures for data input, answering process assistance and visualization of the expert system results.

The range and quantity of control variables give quite a high number of possible algorithm control states, i.e. approx. 200. The expert system determines the method control (which is varied during the analysis) so that the solution be time - optimum and satisfactory for the user.

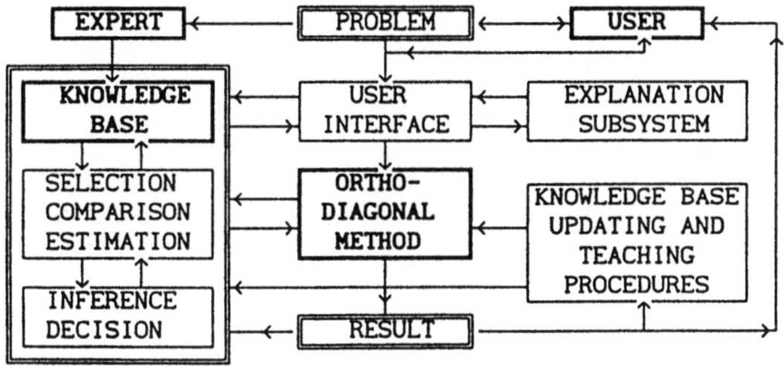

Fig.1. Expert system components

An important factor to be taken into account in the expert systems is the analysis time. Therefore, controlling the optimization analysis should restrict the number of computations of time - consuming elements (e.g. values of objective function) in given method.

The expert determines the way of moving within the decision variables space with an option to omit steps that are needless at given state of the analysis, according to his experience. The declared computation time, type of solution and the analysis of initial course of the task are based on the prepared by the expert knowledge base and influence both the course and computation time of the optimization task.

In continuous development of the knowledge base, an important element is the possibility of making use of completed optimization analyses. The prepared procedures for the information selection and reduction as well as for the inference and decision - making allow the system knowledge base to "self-teach" and to be updated. One of the tasks to be performed by the teaching procedures is to select, basing on a limited initial course of the optimization task, a further course of the task that is most probable for given class of solutions and, next, to accept it. A measure of division into control classes is the dimension of the decision space determined by number of decision variables.

At this definited Euklidian space with the discrete elements, the weight quadratic loss function R is determined. The estimator minimizing the loss function at the initial values of the solution path of the given control S^i was assumed.

The optimization of controls selection using the O-D method is performed so that the loss function $\mathcal{R}$ will be the least. The loss $\mathcal{R}$ is associated wih functional relationship between the erratic classification of initial control S^i and the control S^* that exists in the knowledge base. The loss functions were assumed basing on computed values of the objective functions $f_j^i(x)$ and $f_j^*(x)$ [7]

$$\mathcal{R}(S^i,S^*) = \min_i \sum_{j=1}^{q} \alpha_j \left(f_j^i(x) - f_j^*(x)\right)^2 \ , \ \ i=(1,2,\ldots,n). \tag{1}$$

Introduced weight coefficients assume their maximum values in the final phase of the analysis by the O-D method, for the sake of their decisive significance to the final result.

The selection of that control to which a higher probability of correct classification $\mathcal{P}$ can be attributed was based on decision procedures that employ the Bayes' rules [1,5,10]. The quality of the classifier is considered on assumption that there are two classes: S^i and S^* of the "a priori" probabilites (P^i and P^*) of which are equal. On account of a finite teaching set that results from finite quantity of calculations of the objective function values, the conditional probabilities in these rules were replaced with frequencies.   The images f are one-dimensional and can assume discrete values $f_j(x)$, and, moreover, every conditional discrete distribution $P(f_j(x)|S^i)$ is equiprobable

$$\mathcal{P}(q,S^i,S^*) = 1 - \sum_{j=1}^{q} \max_{(i,*)} \left(P(f_j^i(x)|S^i)\cdot P(S^i)\right) \tag{2}$$

$$P\left(f_j^i(x)|S^i\right) \simeq s^i/m^i, \quad P\left(f_j^i(x)|S^*\right) \simeq s^*/m^*, \quad P(S^i) + P(S^*) = 1 \ , \tag{3}$$

where: m^i, m^* designate numbers of elements in classes S^i and S^*,
$\quad\quad$ s^i, s^* are appearance numbers for values $f_j^i(x)$, in classes S^i
$\quad\quad\quad$ and S^* respectively,
$\quad\quad$ j=1,2,....,q, q - number of objective function computations.

To determine the number of elements of the control class the threshold value d_0 distinguishing the values of the function $f_j^i(x)$ was assumed. It was assumed that two points $f_j^i(x)$ and $f_k^i(x)$ are nondistinguished if the measure d is smaller than value d_0

$$d\left(f_j^i(\mathbf{x}) - f_k^i(\mathbf{x})\right) < d_o \ , \quad k \neq j. \qquad (4)$$

The value d_o can be determined by the expert system depending on the average distance within the control and on the declared by the user accuracy of the solution.

Restarted analysis follows that track which, after the initial analysis, provides the highest probability of a correct classification $\mathcal{P}$. As the entire task is completed, the probability of a correct classification and the loss function are computed. Acceptance of given control to the preferable controls set of the knowledge base takes place when this control has been classified, by the decision rules, as being new or belonging to given class of controls. In such a case, also the number of statistic elements of control models for various optimization tasks is increased.

3. Subject of Optimization

The subject of analysis is an exhibition hall the dimensions of which, in horizontal plane, are 48 m x 48 m [6]. The roof of the hall is formed by a double -layer spatial truss (Fig.2). The truss is supported by pin-joint bearings placed, at 12.0 m intervals, in the upper layer joints. The truss members were designed as made of hot-rolled steel pipes (Table 1).

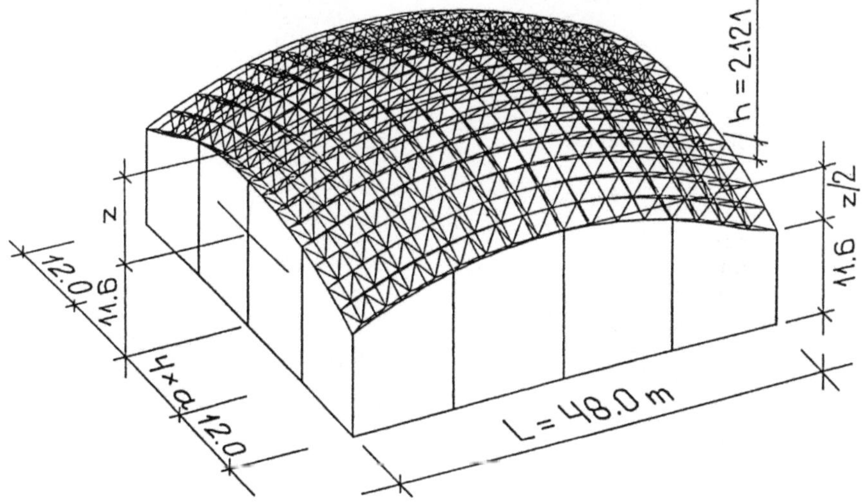

Fig.2. Layout of covering of exibition hall

In designing the truss members the following loads were considered: the weight of truss members and joints, the weight of roof covering and purlins, the snow load and the technological load.

The optimization analysis was carried out for the decision variables vector composed of three elements: $\mathbf{x} = [s, \ z, \ t]^T$, where the form of roofing $s \in \{1,2,\ldots,5\}$, the camber $z \in \langle 0.5, \ 12.0 \ \mathrm{m}\rangle$ and the size of members cross-section cataloque $t \in \{5,9,17,33\}$. The considered forms of roofing : $s \in \{1,2,\ldots,5\}$ are shown in Fig.3.

Table 1

D [mm]	31.8	33.7	38.0	42.4	44.5	48.3	51.0	54.0	57.0	57.0	60.3	63.5	70.0	76.1	76.1	88.9	101.6	101.6	108.0	114.3	133.0	139.7	139.7	159.0	159.0	168.3	168.3	193.7	193.7	219.1	219.3	244.5	244.5
g [mm]	2.9	2.9	2.9	2.9	2.9	2.9	2.9	2.9	2.9	3.2	3.2	3.2	3.2	3.2	3.6	3.6	3.6	4.0	4.0	4.0	4.0	4.0	4.5	4.5	5.0	5.0	5.6	5.6	6.3	6.3	7.1	7.1	8.0
T(33)	1	2	3	4	5	6	7	8	9	10	11	12	13	14	15	16	17	18	19	20	21	22	23	24	25	26	27	28	29	30	31	32	33
T(17)	1		2		3		4		5		6		7		8		9		10		11		12		13		14		15		16		17
T(9)	1				2				3				4				5				6				7				8				9
T(5)	1								2								3								4								5
T(3)	1																2																3

The feasible domain of the task was determined assuming constrains resulting from design-related, technological and computational conditions [4].

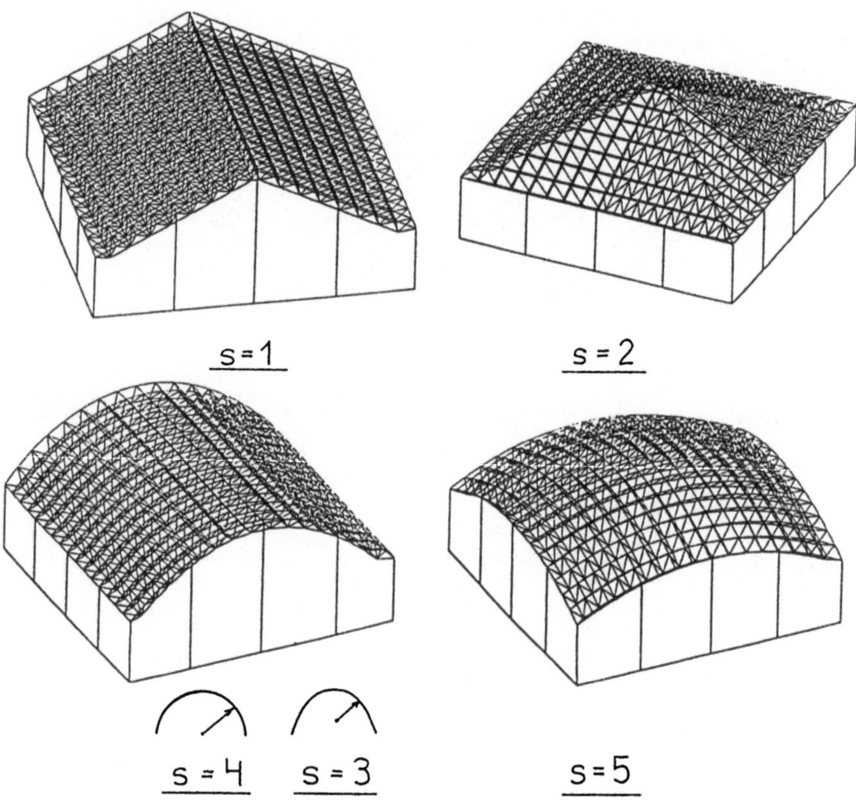

Fig.3. The considered forms of roofing

4. Optimization Analysis

The above formulated problem of selection of spatial truss geometry is of discrete-continuous nature. The decision variables s and t are typical discrete variables while the variable t is of continuous nature. It has been found that an increment in the value of t is equal $\Delta z = 1.2$ m basing on initial analysis of sensitivity of the objective function elements. In the further course of the analysis the variable z is considered as being discrete.

The cross-sections of the truss members are selected automatically by a computational program from the declared series of types. In designing, the local stability was taken into account.

The minimum of the following objective functions are computed:

1° minimum mass of steel elements per 1 m2 of the horizontal projection

$$f_m(\mathbf{x}) = \frac{\gamma}{L^2} \sum_{k=1}^{Lp} L_k \cdot A_k,$$ (5)

2° the rate of utilization of the cross-section of the members

$$f_p(\mathbf{x}) = \frac{1}{Lp} \sum_{k=1}^{Lp} \left| \frac{P_k^{kr} - P_k}{P_k^{kr}} \right|,$$ (6)

Table 2

	S^i	q	$\mathcal{R}(S^i,S^*)$ $\alpha_j = 1$	$\mathcal{R}(S^i,S^*)$ $\alpha_j < 1$	q	m^*	m^i	$\mathcal{P}(q,S^i,S^*)$
$t_{fix}=9$ f_m	S^1	4	0.002	0.000	4	5	5	0.400
	S^1	11	0.006	– – –	9	12	11	0.958
	S^2	5	0.058	0.005	6	5	5	0.100
	S^2	11	0.109	– – –	11	12	11	0.390
$t_{fix}=9$ f_p	S^1	10	0.048	0.002	11	10	10	0.400
	S^1	20	0.048	– – –	16	20	20	0.425
	S^2	10	0.273	0.019	11	10	10	0.200
	S^2	18	0.680	– – –	17	20	18	0.419
$s_{fix}=4$ f_m	S^1	5	0.074	0.008	7	5	5	0.100
	S^1	10	0.075	– – –	11	10	13	0.319
	S^2	5	0.244	0.023	8	5	5	0.000
	S^2	11	0.027	– – –	13	10	8	0.150
$s_{fix}=4$ f_p	S^1	5	0.619	0.068	10	5	5	0.000
	S^1	9	0.734	– – –	13	11	9	0.273
	S^2	5	7.908	0.846	7	5	5	0.001
	S^2	11	8.462	– – –	16	11	13	0.276

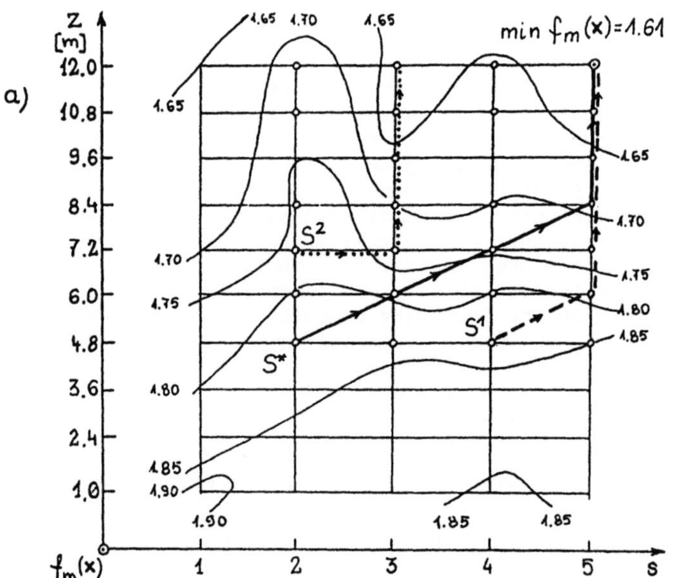

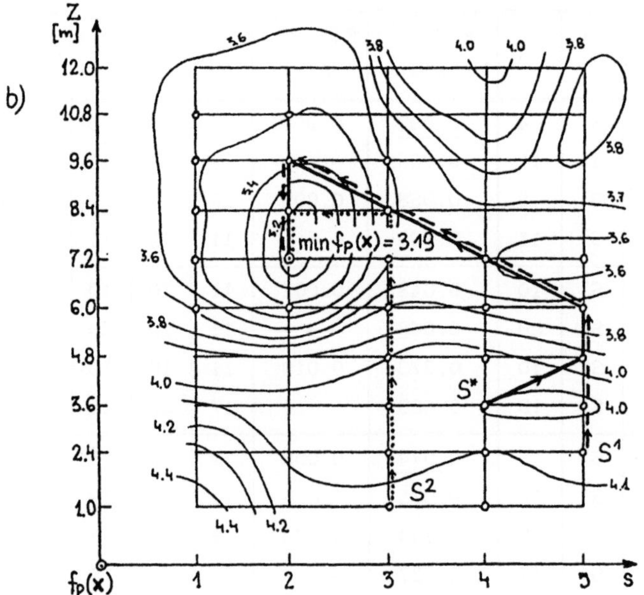

Fig.4. The tracks S^i of moving for the fixed $t=9$

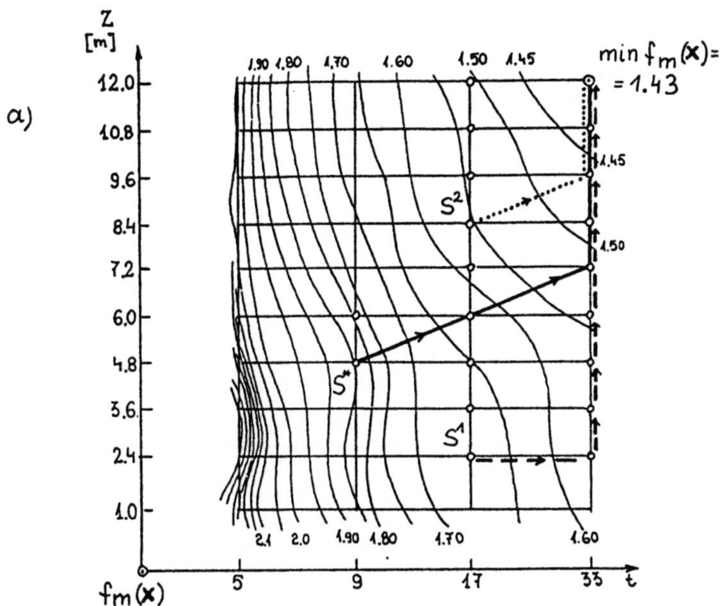

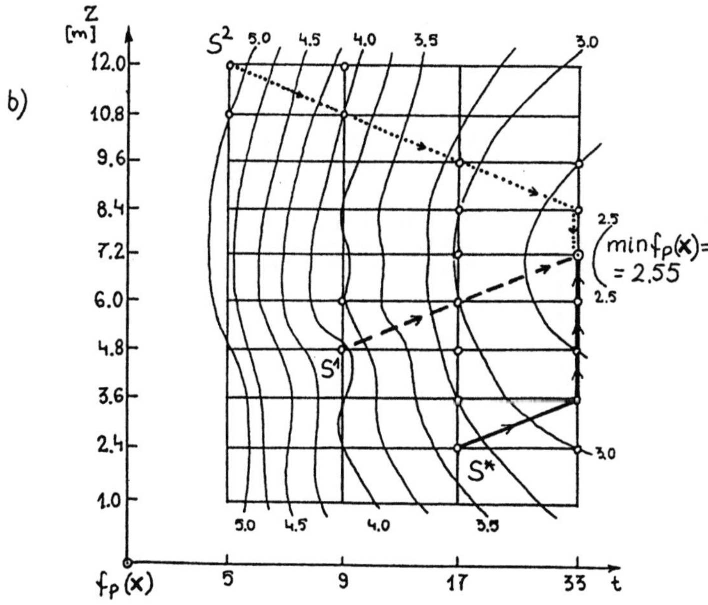

Fig.5. The tracks S^1 of moving for the fixed $s=4$

where:

γ - mass of density of steel, γ = 7850 kg/m^3,
l_k - lenght of k-th member of truss,
A_k - cross-section area of k-th member of truss,
L_p - number of truss members, $L_p \in$ <128, 1152>,
P_k - force in k-th member of truss,
P^{kr} - critical force in k-th member.

Optimization was performed independently for of the functions $f_m(x)$ and $f_P(x)$. Figure 4 and 5 show the tracks S^1 of moving within the feasible domain of the task with the use of the O-D method, at different starting points. The selection of solution track carried out using the knowledge base selection-and-comparison procedures is presented for the initial control S^*. The computed values of the loss function $\mathcal{R}(S^1, S^*)$ for particular controls and the probability of correct classification $\mathcal{P}_i$ allocated to them are shown in Table 2. For

Table 3

	z	s	t	f_m	f_P
	m	-	-	kg/m^2	kN
t_{fix}	12,0	5	9	<u>1.61</u>	3.80
	7.2	2	9	1.76	<u>3.19</u>
s_{fix}	12.0	4	33	<u>1.43</u>	2.91
	7.2	4	33	1.52	<u>2.55</u>

the objective functions $f_m(x)$ and $f_P(x)$, a further course of the analysis was assumed for the control S^1. After reaching the minimum objective function, the values of both the loss function and probability of correct classification were recalculated for entire course of the task (Table 2).

Minimum of f_m and f_P was determined for the variables z and s, with fixed size of the cross-sections catalogue of $t=9$ being assumed (fig.4) and for the variables z and t with fixed type of roofing of $s=4$ being assumed (fig.5).

The preferable vector of decision variables is shown in Table 3. The preferable solutions for fixed size of the cross-sections are a double-curvature truss for function f_m and a four-sloping truss for f_P. When form of roofing is a cylindrical roof, then the preferable cambers of the crowning are equal 12 m for f_m and 7.2 m for f_P.

5. Conclusions

The choice of the optimization path based on the analysis of the initial path in the decision variables space can be done by the means of the comparison and selection procedures introduced in the expert system. It decreases the search time necessery to find the minimum of the objective function, which is of great significance in the case of large structures.

The computer program OPTIM based on the Ortho-Diagonal Method of the discrete optimization makes an effective instrument for the structural designing. The expert system controlling the process of the analysis enables satisfactory solution to be obtained in time acceptable for user.

The expert system used in the OPTIM helps a user to formulate tasks of discrete optimization. It is active during the analysis and influences on the time and character of solutions obtained. In sets of results it informs on the analysis and the solution. The OPTIM in many cases can help in optimum discrete designing.

References

1. Devijver P., Kittler J. Pattern Recognition: A Statistical Approach. Prentice Hall, London 1982.
2. Frost R.A. Introduction to knowledge base system. McGraw-Hill, New York, 1986.
3. Hurley R.B. Decision Tables in Software Engineering. Van Nostrand Reinhold Company Inc., New York, 1983.
4. Karczewski J.A., Niczyj J., Paczkowski W.M. Discrete Optimization of the Space Truss being Controlled by An Expert System. IASS Symp. Spatial Structures at the Turn of the Millenium, Copenhagen, Denmark, 1991, Vol.III, pp.191-197.
5. Krawczyk H., Malina W. Learning to Improve System Level Fault Diagnostic. Proc. XI Conf. on Fault Tolerant System Diagnosis, Vol.III, Suhl, Germany, pp.265-271.
6. Karczewski J.A., Niczyj J., Paczkowski W.M. Multicriterion Selection of Spatial Truss Crown. Proceedings, IASS-Symposium, Istanbul, Turkiye, 1993, pp.461-470.
7. Lehmann E.L. Theory of point estimation. John Wiley and Sons, Inc., London, 1984.
8. Paczkowski W.M. Discrete Optimization Program Based on The Ortho-Diagonal Method, Int. Conf. on Optimization Techiques and Applications (ICOTA'92), Singapore, 1992, Vol.1, pp.248-255.
9. Shoup T.E., Misttree F. Optimization Methods with Applications for Personal Computers, Prentice-Hall Inc., New Jersey, 1987.
10. Sas J. Probabilistic Approach to Pattern Recognition in Expert Systems. Proc. 7th Int. Conf. on Systems Engineering, Las Vegas 1990.

Discrete Structural Optimization Based on Multipoint Explicit Approximations

Vassili V. Toropov[1], Valery L. Markin[2], Henrik Carlsen[3]

[1] Department of Civil Engineering, University of Bradford, Bradford, West Yorkshire, BD7 1DP, UK

[2] Laboratory for Engineering Mechanics, Delft University of Technology, P.O. Box 5033, 2600 GA Delft, The Netherlands

[3] Laboratory for Energetics, Building 403, The Technical University of Denmark, Lyngby, DK-2800, Denmark

Abstract. Many real-life structural optimization problems have the following characteristic features: (i) some or all design variables can be defined as a set of discrete parameters, (ii) the objective function and constraints are implicit functions of design variables, (iii) to calculate values of these functions means to use some numerical response analysis technique which usually involves a large amount of computer time, (iv) the function values and (or) their derivatives often contain noise, i.e. can only be estimated with a finite accuracy. To cope with above problems, the *multipoint approximation method* has been developed. It is considered as a general iterative technique, which uses in each iteration simplified approximations of the original objective/constraint functions. They are obtained by the multiple regression analysis methods which can use information being more or less inaccurate. The technique allows to use in each iteration the information gained in several previous design points which are considered as a current design of numerical experiments defined on a given discrete set of parameters. It allows to consider instead of the initial discrete optimization problem a sequence of simpler mathematical programming problems and to reduce the total number of time-consuming numerical analyses. The obtained approximations are considered to be valid within a current subregion of the space of design variables defined by discrete values of move limits. This approach provides flexibility in choosing design variables and objective/constraint functions and allows the designer to use his experience and judgment in directing the optimization process.

1 Formulation of a discrete-continuous optimization problem

Consider a general discrete-continuous structural optimization problem:

Minimize

$$F_0(x) \tag{1}$$

subject to

$$F_j(x) \le C_j \, , \, (j = 1,..., M) \tag{2}$$

and

$$A_i \leq x_i \leq B_i, \quad (i = 1,..., N) \tag{3}$$

where

$x = (x_C, x_D)$ is a vector of design variables which can be varied in the course of the design procedure, where the subvector $x_C = (x_1,..., x_{NC})$, $x_C \in R^{NC}$ presents continuous design variables and the subvector x_D presents the discrete values of ND design variables chosen from the given finite set: $x_{Di} = (x_i^1, x_i^2,..., x_i^{n_i})$, $i = 1,..., ND$. Without imposing restrictions of the generality of our considerations, we assume that $x_i^{j+1} > x_i^j$, $i = 1,..., ND$, $j = 1,..., n_i -1$. The total number of design variables $N = NC + ND$.

$F_o(x)$ is an objective function which provides a basis for choice between alternative acceptable designs;

$F_j(x)$, $(j = 1,..., M)$ are the constraint functions which impose limitations on various behavioral characteristics of the structure;

A_i and B_i are side constraints, describing physical upper and lower limitations on design variables. Note that for the discrete variables $A_i = x_i^1$ and $B_i = x_i^{n_i}$, $i = NC+1,..., N$.

Depending on the specific problem under consideration, functions $F_j(x)$, $(j = 0,..., M)$ can describe various structural response quantities.

2 Multipoint approximation technique

2.1 Introduction

The formulated optimization problem has the following characteristic features:

- some or all design variables can be defined as a set of discrete parameters;

- the objective function and constraints are implicit functions of design variables;

- to calculate values of these functions means to use some numerical response analysis technique which usually involves a large amount of computer time;

- function values and (or) their derivatives often contain noise, i.e. can only be estimated with a finite accuracy.

The prohibitive computational cost of the direct combination of numerical analysis programs with conventional methods of mathematical programming stimulated the idea of sequential approximation of the initial optimization problem (1) - (3) by explicit subproblems. Nowadays this concept has proven to be very efficient and it is widely used in structural optimization (see a recent review [1]). It is based on the information obtained by numerical response analysis and first order design sensitivity analysis (i.e. values of functions and their derivatives) at a current point of the design variable space (*single point approximations*). At present, several first order approximation techniques are known, which are based upon the function value and its derivatives at the current and the last previous design points (*two-point approximations*) [1]. The drawback of these approaches is that the convergence of optimization algorithms can be very slow if the objective/constraint function values present a considerable level of noise. Moreover, quite often it is very difficult (or even impossible) to obtain the values of derivatives of the objective and/or constraint functions with respect to discrete design varibles.

A different approach to complex engineering systems optimization [2], [3] is to create approximate explicit expressions by analysing a chosen set of design points (design of experiments). This approach is based on the multiple regression analysis methods which can use information being more or less inaccurate. They are global in nature and allow to construct explicit approximations valid in the entire design space. The numerical experiments are performed at a pre-selected set of discrete points which can be allocated at individual points of a grid of discrete design variables. The drawback of this approach is that it is restricted by relatively small optimization problems (up to ten design variables [2]).

The remainder of this section presents a general iterative technique which has been used for various complex engineering systems optimization problems. The simplified approximations of the original objective/constraint functions are obtained by the weighted least-squares method. It uses in each iteration the information about function values gained in several previous design points (multipoint approximations [4]) which are considered as a current design of numerical experiments defined on a given discrete set of parameters. It allows to consider instead of the initial discrete optimization problem a sequence of simpler mathematical programming problems and to reduce the total number of time-consuming numerical analyses. We believe that it combines the advantages of two above mentioned basic approaches.

2.2 Basic approximation concept

The approximation concept leads to the iterative approximation of the original functions $F_j(x)$, $(j = 1,..., M)$ by the simplified functions $\tilde{F}_j(x)$. The initial optimization problem (1) - (3) is replaced with the succession of simpler mathematical programming subproblems as follows:

Find the vector x_*^k that maximizes the objective function

$$\tilde{F}_o^k(x) \tag{4}$$

subject to

$$\tilde{F}_j^k(x) \leq C_j \, , \, (j = 1,..., M) \tag{5}$$

and

$$A_i^k \leq x_i \leq B_i^k, \quad A_i^k \geq A_i, \quad B_i^k \leq B_i, \quad (i = 1,..., N), \qquad (6)$$

where k is the current iteration number. Note that the current move limits A_i^k and B_i^k define a subregion of the design space where the simplified functions $\tilde{F}_j^k(x)$ can be considered as adequate approximations of the initial functions $F_j(x)$. In the case of discrete design variables, the move limits correspond to nodes of a grid of these variables..To estimate the order of the adequacy of simplified functions in comparison with the initial functions, the error parameters

$$r_j^k = |[\tilde{F}_j^k(x_*^k) - F_j^k(x_*^k)]/F_j^k(x_*^k)|, \quad (j = 0, 1,..., M) \qquad (7)$$

can be evaluated.

Various discrete (or continuous-discrete) optimization techniques can be used to solve the subproblem (4) - (6) because these functions are very simple.

The next, $(k+1)$-th iteration is started from the obtained point x_*^k.

2.3 Multipoint approximations

To construct the simplified expression $\tilde{F}_j^k(x)$ in the subproblem (4) - (6), we shall implement the methods of regression analysis [5]. These methods are intended for obtaining an expression that reflects the behaviour of an object considered as a function of its parameters, based on a discrete set of experimental results. Here and in the remainder of this section, *an experiment* means a numerical experiment using some numerical response analysis. It is essential to note that we do not intend to construct simplified expressions that are adequate in the whole of the search region determined by side constraints A_i and B_i in (3) because it takes too large number of numerical experiments in the case of a multiparameter problem. Therefore, we construct such expressions iteratively only for separate search subregions which are determined by move limits A_i^k and B_i^k at each k-th step of the problem in (4) - (6). Thus, the functions $\tilde{F}_j^k(x)$, $(j = 0,..., M)$ give piece-wise approximation of the initial functions $F_j(x)$.

To simplify notation, we will suppress the indices k and j on the functions $\tilde{F}_j^k(x)$. Assume that the functions (4), (5) are expressed in the following general form:

$$\tilde{F} = \tilde{F}(x, a) \qquad (8)$$

The vector $a = (a_0, a_1,..., a_L)^\mathsf{T}$ in expression (8) consists of so-called tuning parameters, that is, free parameters the value of which is determined on the basis of numerical experiments at points located at several individual nodes of a grid of design variables. Then the weighted least-squares method leads to the following problem:

Find the vector a that minimizes the function

$$G(a) = \sum_{p=1}^{P} w_p \, [\, F(x_p) - \tilde{F}(x_p, a)\,]^2, \tag{9}$$

where p is the number of a current point in the plan of experiments, P is the total number of such points, w_p is the weight coefficient that characterizes the relative contribution of the p-th experiment information. The solution of the optimization problem (9) is the vector a which makes up the simplified function (8).

Let us formulate the problem of vector a evaluation if in addition to values of function $F(x_p)$ their first order derivatives $F(x)_{,i} = \partial F(x) / \partial x_i$, $(i = 1, ..., N)$ at points x_p, $(p = 1,..., P)$ are known.

In this case we shall minimize the function

$$G(a) = \sum_{p=1}^{P} \{ w_p^{(0)} \, [F(x_p) - \tilde{F}(x_p, a)]^2 + \sum_{i=1}^{N} w_p^{(i)} [F(x_p)_{,i} - \tilde{F}(x_p, a)_{,i}]^2 \} \, , \tag{10}$$

where $w_p^{(0)}$ and $w_p^{(i)}$ are the weight coefficients which characterize the relative contribution of information about $F(x_p)$ and $F(x_p)_{,i}$ correspondingly in the p-th experiment. *An experiment* means in this case a numerical analysis analysis including the design sensitivity analysis. Note that the formula (10) could easily be extended if higher order sensitivities would become available.

The quality of approximations depends strongly on values of weight coefficients $w_p^{(i)}$ in (10), they reflect the inequality of data obtained in different design points. Usually, the point that corresponds to the solution of a complex engineering systems optimization problem (1) - (3) lies on a boundary of the feasible region. Therefore, the quality of approximations should not be the same within the current search subregion. On the contrary, it should be the best in a domain, which is located along the boundary of the feasible region. This is possible if the weights depend on the distance of a current point from the boundary of the feasible region, i.e. on the values of $F(x)$. Similarly, weights can reflect the difference in the contribution of the information given by different experiments depending on the objective function value at individual design points. Then the maximum value of the weight coefficient corresponds to the design point with the maximum value of the objective function.

2.4 Choice of the simplified expression structure

To construct the simplified expressions $\tilde{F}$, it is necessary to define them as a function of tuning parameters a . Apparently, the efficiency of the optimization technique depends greatly on the accuracy of such expressions. The simplest case is a linear function of parameters a :

$$\tilde{F}(a) = a_0 + \sum_{l=1}^{L} a_l \, \varphi_l \, . \tag{11}$$

If the results $F(x_p)$, $(p = 1,...,P)$, of P numerical experiments are known, then the minimization problem in (9) or (10) is equivalent to the solving of the linear system of $L+1$ normal equations with $L+1$ unknowns a_l:

$$[\Phi]^\mathsf{T} [W] [\Phi] \, a = [\Phi]^\mathsf{T} [W] \, f, \tag{12}$$

where $[\Phi]$ is the rectangular matrix consisting of values of individual functions φ_l (regressors) from the expression (11) at every plan point; f is the vector containing information from experiments obtained at P plan points, i.e. values of implicit functions $F(x_p)$; $[W]$ is the diagonal matrix consisting of weight coefficients w_p.

Note that the structure of the simplified expression (11) is rather general because the individual regressors φ_l can be arbitrary functions of design variables.

The procedure described above can be further generalized by the application of intrinsically linear functions [5]. Such functions are nonlinear, but they can be led to linear ones by simple transformations. For example, the multiplicative function

$$\tilde{F}(a) = a_0 \, \varphi_1{}^{a_1} .... \, \varphi_L{}^{a_L} \tag{13}$$

requires the logarithmic transformation

$$\ln \tilde{F}(a) = \ln a_0 + \sum_{l=1}^{L} a_l \ln \varphi_l . \tag{14}$$

Several particular forms of intrinsically linear functions were implemented for the solution of various complex engineering systems optimization problems [6]. The obtained results indicate that the multiplicative function (13) is the most universal one. It should be noted, however, that the structure of the simplified function need not necessarily be linear (as in (11)) or intrinsically linear (as in (13)). If there are arguments for the application of nonlinear expressions in the general form, then the problem of the least-square estimation (9) or (10) would become a standard problem of nonlinear programming, which can be solved by any available technique. Moreover, the simplified functions $\tilde{F}$ need not necessarily be explicit. There can be numerical procedures involved in their formulation, such as numerical integration. But, the basic requirements to such simplified models are:

- its description of the simplified process must depend on the same parameters x as the initial numerical model, presented by functions $F(x)$;

- they have to contain some tuning parameters a, which are obtained by solving the optimization problem (9) or (10);

- they have to be simple enough to be used in numerous repeating calculations;

- in order to achieve fast convergence of the algorithm, they have to be accurate enough in comparison with the original functions (1), (2);

- they have to be noiseless or, at least, the level of noise must not cause problems with convergence of an algorithm used to solve the optimization problem (4) - (6).

2.5 Choice of a search subregion

After formulation of the simplified functions (8), the current mathematical programming problem in (4) - (6) is solved and the error parameters r_j^k in (7) for the point x_*^k are estimated.

Then the task is to determine the move limits A_i^k and B_i^k for the next iteration. First, the condition

$$r_j^k \le e_j^k, \quad (j = 0,...,M) \tag{15}$$

is checked, where e_j^k are small positive values which define the feasible accuracy of approximation of functions $F(x)$ by functions $\widetilde{F}_j^k(x)$. If this condition is not satisfied even for one of the active constraints or objective function, (i.e. approximations are inaccurate), then the size of the search subregion of the $(k+1)$-th step must be reduced. When the conditions (15) are satisfied for all active constraints and the objective function, we must decide upon the movement of the search subregion. If the point obtained, x_*^k, is located inside the k-th search subregion (none of the move limits are active), then this point can be considered as the current approximation of the solution x_*. In that case, the next search subregion should be reduced and the other conditions of the search termination should be checked. Otherwise, the search must be continued. This means that the search subregion must be moved in the direction $x_*^k - x_*^{k-1}$. The search process is terminated when (i) the conditions (15) are satisfied, (ii) none of move limits are active and (iii) the subregion has reached a required small size.

3 Example: Optimization of a Stirling engine

The above described optimization technique was used for the performance optimization of small Stirling engines developed at the Technical University of Denmark. They are designed to be used as individual combined heat and power plants (co-generation plants). Such engines driven from natural gas or biomass can supply a single family house with heat and power with a high level of energy utilization. It is considered a one of possible ways of decentalization of energy supply in rural areas.

The basic Stirling engine working cycle (Fig. 1) is a closed cycle, where the working gas is kept inside the cylinders, and heat is added and removed from the working space through heat exchangers. The Stirling engine differs therefore from conventional internal combustion engines by having external combustion, like in a boiler (Fig.2). All sorts of gas, liquid and solid fuels can be used depending on the proper external burner system only.

The ideal Stirling working cycle has the maximum obtainable efficiency defined by Carnot efficiency, and highly efficient Stirling engines can therefore be built, if designed properly. Heat exchangers, regenerators and cylinder volumes have to be optimized carefully for heat transfer, flow losses, dead volume, etc. To analyse the power output and the efficiency of a Stirling engine under consideration, numerical simulation programs (NSP) have been developed, which calculate power output, efficiency and heat flow from the thermodynamic equations. These differential equations are solved by means of the fourth order Runge-Kutta method. The initial pressure and temperatures for steady state operation are found by iterative procedure. To verify NSP, the numerical results have been compared to the results of laboratory experiments with a 10 kW natural gas driven Stirling engine. The comparison showed that NSP allows to predict the Stirling engine performance quite accurately.

To describe the engine design realistically for the optimization purposes, it is necessary to consider several tens of design variables. Some of these design parameters describe a number of individual parts of the engine, therefore, they have the discrete nature. Furthermore, objective and constraint functions of the optimization problem present some level of noise, i.e. can only be estimated with a finite accuracy due to the iterative nature of NSP.

The efficiency has been maximized, while the power output has been controlled by two constraints. The first constraint forces the power output not to be less than 10 kW, and the second constraint keeps the specific power above a certain minimum. Other constraints ensure the strength characteristics of the engine components.

The objective and constraint functions have been approximated by multiplicative expressions (12) in the simplest form:

$$\tilde{F}(a) \;=\; a_0\, x_1{}^{a_1}\ldots x_{18}{}^{a_{18}}.$$

Table 1 shows the results of the optimization of the Stirling engine, which has been used for tests in the laboratory. The optimization was carried out using 18 design parameters, the results showed that the efficiency increased from 30.1% to 41.2% with the same power output without changing the important design parameters like bore and stroke by more than 20%. Solutions were found which had never occurred by the traditional "trial and error" approach. Also the optimization helped the designer to understand the fundamentals of Stirling engine performance in a completely new way, and the possibilities for improvements were found to be numerous.
The number of calls for the numerical simulation program in all runs has not been greater than 200.

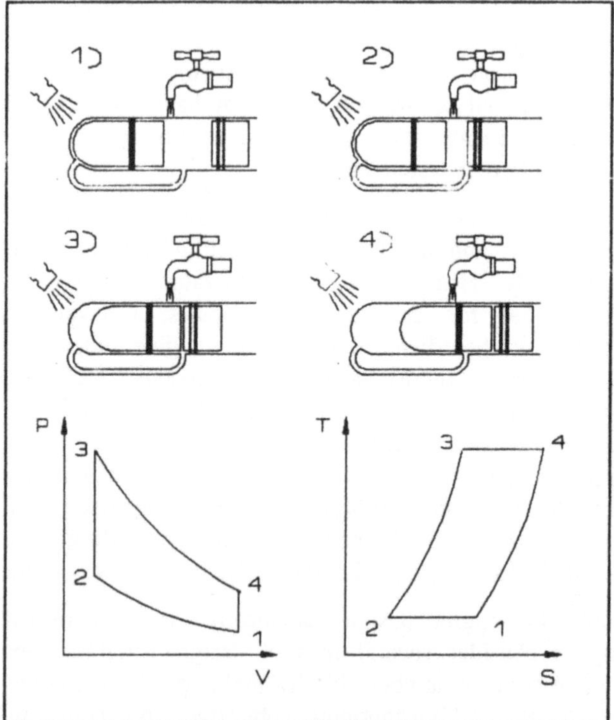

Fig. 1. Stirling engine working principle

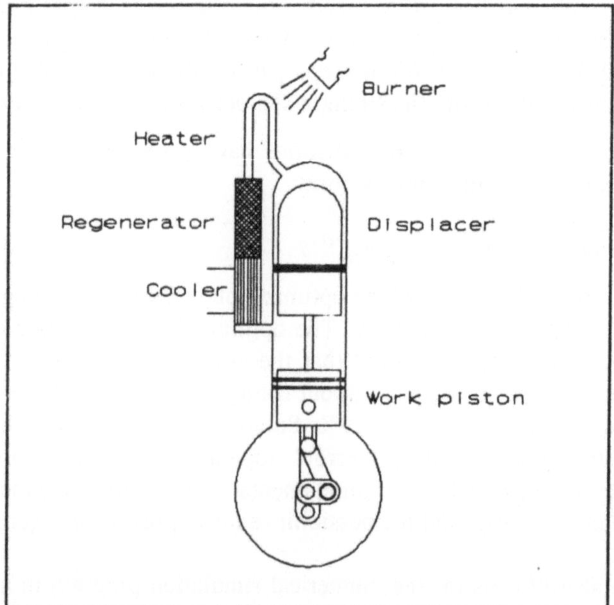

Fig. 2. Schematic Stirling engine

Table 1. Results of the optimization process

No	Design variable	Unit	Starting value	Optimum value
1	Bore, displacer piston	mm	105	121
2	Bore, power piston	mm	95	125
3	Stroke, displacer piston	mm	51	63
4	Stroke, power piston	mm	60	63
5	Phase angle	deg.	70	66.8
6	Mean pressure	MPa	8	4.73
7	Speed	rpm	1550	1050
8	Displacer gap	mm	0.75	0.7
9	Regenerator with	mm	25.4	39.2
10	Regenerator length	mm	44	66
11	Regenerator filler factor		0.22	0.224
12	Regenerator thread diameter	mm	0.04	0.71
13	Number of heater tubes		24	45
14	Number of cooler tubes		216	270
15	Length of heater tubes	mm	365	477
16	Length of cooler tubes	mm	180	146
17	Internal diameter of heater tubes	mm	8	5.1
18	Internal diameter of cooler tubes	mm	2.5	2.68
EFFICIENCY		%	30.1	40.2
POWER OUTPUT		kW	10.1	10.0

References

1. Barthelemy, J.-F.M, Haftka, R.T.: Approximation concepts for optimum structural design - A review. *Structural Optimization* **5** (1993), 129-144

2. Vanderplaats, G.N.: Effective use of numerical optimization in structural design. In: *Finite Elements in Analysis and Design* **6** (1989), 97-112, Elsevier

3. Schoofs, A.J.G.: Experimental design and structural optimization. Ph.D. Thesis, Eindhoven University of Technology, The Netherlands, 1987

4. Toropov, V.V.: Multipoint approximatiom method in optimization problems with expensive function values. *Computational Systems Analysis* 1992, 207-212, 1992, Elsevier

5. Draper, N.R.; Smith, H. : Applied regression analysis. 2nd ed. N.Y.: Wiley, 1981

6. Toropov, V.V.; Filatov, A.A.; Polynkin, A.A: Multiparameter structural optimization using FEM and multipoint explicit approximations. *Structural Optimization* **6** (1993), 7-14

Optimization of Perturbation Parameters for Forced Vibration Stress Levels of Turbomachinery Blade Assemblies

E.Petrov
Kharkov Polytechnic Institute
Ul. (*street*) Frunze 21, Kharkov, 310002, UKRAINE

Keywords. Turbomachinery, blade, perturbation, vibration.

1. Introduction

Connecting turbomachinery rotor blades by interblade linkage is a widely used technique of boosting reliability and strength of axial flow impellers of gas engines and power generating turbines. Stress levels from forced vibrations are highly sensitive to small deviations in blade geometry and the dimensions of the blade-shroud assembly. Deviations by 1-2% may lead to a change of stress levels by a factor 1.5 to 2.0. Therefore knowledge of most favorable and most disadvantageous deviations in sets of blades, such as may occur within manufacture tolerances, is of importance.

Influence of blade geometry and dimensions on amplitudes of forced vibrations of turbomachine impellers with an interblade linkage and a stiff disc is considered in this paper. A technique is put forward to calculate either the optimal or the most adversely disadvantageous sets of blade perturbation within existing manufacture tolerances, from the point of view of stress levels. A technique is developed of forced vibration calculation enabling to study systems with an arbitrary number of blades and with any distribution and magnitude of the imposed deviations (parameter perturbations or mistuning). Any design of blade linkage as used in turbomachinery may be treated, say, e.g. circular, packet, "chess bond" and others, including such interconnection designs that disturb the cyclic symmetry of the impeller (Fig. 1a,b).

The coupled bending-torsional-longitudinal vibrations of blades and interblade linkage is studied in the paper and a blade model of a high degree of completeness is used. Inhomogeneity of excitation loads and the fact that these loads may have various phase shifts are taken into account so as also fluid damping, blade interaction through fluid flow, damping in blade material and in blade-shroud joints.

2. Statement of the Optimization Problem

Finding the perturbation of the blade geometry such that maximum or minimum stress amplitudes are reached is an optimization problem. The *objective function* of such a problem is the maximum stress amplitude σ_{max}. These amplitudes are found within the sets of blade system points and of excitation frequencies within a given frequency range. Datum sections used in manufacturing for defining blade shape are used for controlling deviations. The perturbation parameters which characterize size deviations of blade sections from standard design values μ_{jn} are adopted as parameter variables related to the n-th cross section of j-th blade. The constraints imposed are given by extreme deviations from admissible geometrical dimensions within tolerance limits (Fig. 1c).

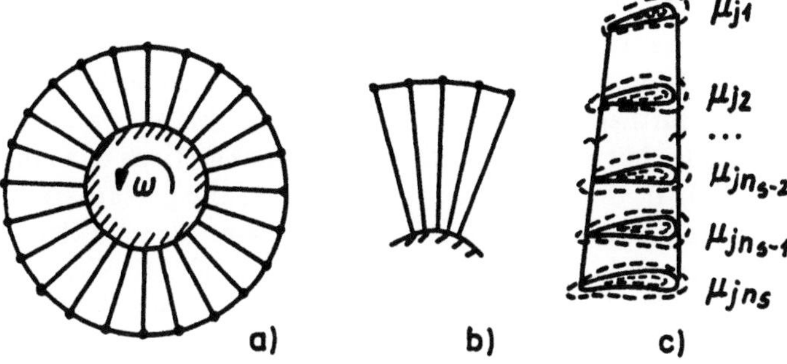

Fig.1. Example of shrouded blade systems (a), (b)
and of vibration of blade geometry (c).

Thus the problem is that of calculating a set of perturbation parameters $\{\mu_{jn}\}$ which make the objective function K attain an extreme value. The objective function is given by $K(\{\mu_{jn}\}) = \sigma_{max} \rightarrow min(max)$ where $\sigma_{max} = \max \sigma(\Omega \in [\Omega^-, \Omega^+]; \quad x, y, z \in V)$ and Ω is the excitation frequency, V - the space region occupied by the impeller blades; Ω^-, Ω^+ are the bounds of admissible excitation frequencies. The constraints on parameter variables may be written: $\mu_{jn}^- \leq \mu_{jn} \leq \mu_{jn}^+$. For solution's sake this constrained optimization problem is transformed into an unconstrained one by the penalty function method. The constraints are incorporated into the objective function which becomes in this case:

$$K(\{\mu_{jn}\}) = \sigma_{max} + g(\{\mu_{jn}\}) \tag{1}$$

where the penalty function is written as

$$g(\{\mu_{jn}\}) = C \sum_{j=1}^{N} \sum_{j=1}^{n_s} |< \mu_{jn} - \mu_{jn}^- >| + |< \mu_{jn}^+ - \mu_{jn} >| \tag{2}$$

where N is the number of blades in the system under scrutiny; n_s is the number of blade datum sections; $<>$ denotes a so called *truncation function*:

$$< \alpha >= \begin{cases} \alpha, & \text{when} \quad \alpha \leq 0 \\ 0, & \text{when} \quad \alpha > 0 \end{cases}$$

C is the penalty coefficient or weight. It is a large positive number for finding a minimum of σ and it is negative for calculating a maximum. A few methods of direct search as in [1] are used.

3. Reducing the Number of Parameter Variables

The number of parameter variables is restricted by requirements of effectiveness of optimization method and computer time consumption needed for computation of forced impeller vibrations for every set of $\{\mu_{jn}\}$. Reducing the number of parameter variables can be performed by assuming that only some of the blade datum sections differ from their standard, the rest being perfect. It may further be assumed that the perturbation parameters differ for different blades but remain the same for various sections of a single blade. Thus $(\mu_{jn} = \mu_j)$. In the latter case the number of the perturbation parameters equals the number of the blades in the system. Usually the number of blades of a system is high, e.g. it is the number of all impeller blades. The need arises to further lower the number of parameter variables. This is done by a Fourier series expansion of the perturbation parameters:

$$\mu(\varphi) = 1 + \sum_{k=1}^{n_p} a_k \sin(k\varphi + \epsilon_k), \tag{3}$$

where φ is the circumferential angle coordinate, n_p is the number of harmonic modes retained. The parameter variables a_k, ϵ_k are n_p in number. The constraints become $\mu^- \leq a_k \leq \mu^+$; $0 \leq \epsilon_k \leq 2\pi$. Thus dimensionality of the optimization problem may be brought down to suit any needs. This impairs the generality of the problem somewhat since in order accurately to represent a set of N numbers such as $\{\mu_j\}$ $N/2$ harmonic nodes are needed. This may however be obviated to some extent by working with selected single harmonic modes or arbitrarily selected blade parts.

4. Calculating the Forced Vibrations

The maximum stress resulting from forced vibration is calculated for a current set of perturbation parameters in every iteration step of the optimization search. The main difficulties encountered when investigating vibrations of blade systems of real complexity are: the large dimensions of matrices of resolving equations, computation complexity and also those difficulties which arise from the necessity to provide accuracy when calculating complex dynamic systems with closely spaced eigenfrequencies. Optimization makes computation speed limitations particularly restrictive. Difficulties encountered determine many details of forced vibration computation technique to a high degree. A combination of the Transfer Matrix Method (TMM), the Finite Element Method (FEM) and a partition into substructures form the basis of the computation technique.

4.1 Excitation of Blade Vibrations

Forced steady state vibrations of blade system are being investigated These vibrations are excited by circumferential and radial inhomogeneous loads appearing during impeller rotations:

$$P(z,\varphi) = \{P_x(z,\varphi), P_y(z,\varphi), P_\psi(z,\varphi)\}^T ,$$

where $P_x(z,\varphi)$, $P_y(z,\varphi)$ are the distributions along a blade of those forces which act in the axial and the circumferential directions respectively; $P_\psi(z,\varphi)$ is the distribution of the torsional moment along the blade. Loads are next expanded into Fourier series. Since the coordinates of vibration reference rotate with the angular velocity ω we have the series:

$$P(z,\varphi,t) = P_o(z) + \sum_{k=1}^{\infty} P_k(z)e^{ik(\varphi+\omega t)} + \overline{P}_k(z)e^{-ik(\varphi+\omega t)} \tag{4}$$

where $P_k(z) = \{P_x^{(k)}(z), P_y^{(k)}(z), P_\psi^{(k)}(z)\}^T$ - column - vector of complex functions which describe the variations of amplitudes and phases of k-th harmonic component along the blade wherein $i = \sqrt{-1}$. Responses are calculated for each harmonic component separately.

4.2 The Matricial Relationship for a Blade

The blade considered is modelled as a twisted variable cross section rod (cf. [2]). Damping in the rod material is introduced by making the moduli of elasticity and of shear assume complex values. Fluid damping and interaction between vibrating blades and the surrounding flow are taken care of by fluid damping influence coefficients (FDIC). Equations of longitudinal, flexural and torsional vibrations at a frequency $\Omega = k\omega$ may be written in the form:

$$[E(Aw' + \tau_o J_p \psi')]'' - \rho\Omega^2 Aw = 0$$

$$[E(J_y u'' + J_{xy} v'' - \tau_o J_{yr}\psi')]'' + \rho\Omega^2 Au = p_x^{(k)}(z) - R_x$$

$$[E(J_{xy} u'' + J_y v'' - \tau_o J_{xr}\psi')]'' + \rho\Omega^2 Av = p_y^{(k)}(z) - R_y \tag{5}$$

$$[GJ_t\psi' + E\tau_o(J_p w' - J_{yr} u'' - J_{xr} v'')]'' - \rho\Omega^2 J_p\psi = p_\psi^{(k)}(z) - R_\psi$$

where $R_\xi = \frac{1}{2}\rho_f v_f^2 c \sum_{j=-n_f}^{n_f} k_{jx}^{(\xi)} u_j + k_{jy}^{(\xi)} v_j + bk_{j\psi}^{(\xi)} \psi_j$ are forces of fluid damping and of interaction with the gas flow (cf. [3]). Here $\xi = x, y, \psi$; $c =$

112

1 when $\xi = x, y$; $c = b$ when $\xi = \psi$; b is the characteristic size of blade profile; ρ is the blade material density; $k_{jx}^{(\xi)}$, $k_{jy}^{(\xi)}$, $k_{j\psi}^{(\xi)}$ is the dimensionless complex FDIC; ρ_f, V_f is the density and velocity of undisturbed flow; n_f is the maximum number difference between blades interacting through the flow; $E = E_0(1 + i\psi_E/2\pi)$ and $G = G_0(1 + i\psi_G/2\pi)$ are the complex moduli of elasticity wherein ψ_E, ψ_G are the damping or dissipation coefficients. Further $A, J_x, J_y, J_{xy}, J_p, J_{xr}, J_{yr}, J_t, \tau_0$ are the area, inertia and torsional stiffness moments of a cross section and the velocity of twist propagation along the blade. The matricial relationship for a blade is calculated by means of the transfer matrix method. In order to perform semianalytical integration of the equations Eq.(5) and to obtain expressions of the transfer matrix entries in the form of definite integrals the blade is partitioned into finite elements with concentrated masses. Large enough number of such elements is chosen so as to provide an accurate description of inertia properties within the whole range of forced vibration frequencies being analyzed, wherein both the low and high natural modes may be appreciable. The transfer matrix of a n-th element of a blade relates vibration parameters at both end sections of such an element:

$$X_{n+1} = L_n X_n + P_n \tag{6}$$

where $X_n = \{u, v, w, \varphi, \theta, \psi, F_x, F_y, F_z, M_x, M_y, M_z\}^T$ is a column vector of complex amplitudes of displacements both linear u, v, w and angular φ, θ, ψ, forces: lateral F_x, F_y and longitudinal F_z, moments: flexural M_x, M_y and torsional M_z. Further L_n is the (12×12) transfer matrix and P_n is the vector of load components.

The recursive property of the relation (6) is conducive to a relation for a whole blade or for a part of it:

$$X_{n_e+1} = \left(\prod_{n=n_e}^{1} L_n \right) X_1 + \sum_{j=1}^{n_e} \left(\prod_{n=n_e-j}^{1} L_n \right) P_j \tag{7}$$

where n_e denotes the number of elements constituting the blade either whole or a part of it.

Eq.(7) is next put into a form which is used in the finite element method:

$$\left\{ \begin{array}{c} Q_1 \\ Q_{n_e+1} \end{array} \right\} = \left[\begin{array}{cc} -l_{12}^{-1}l_{11}, & l_{12}^{-1} \\ l_{21} - l_{22}l_{12}^{-1}l_{11}, & l_{22}l_{12}^{-1} \end{array} \right] \left\{ \begin{array}{c} q_1 \\ q_{n_e+1} \end{array} \right\} + \left\{ \begin{array}{c} -l_{12}^{-1}p_q \\ P_Q - l_{22}l_{12}^{-1}p_q \end{array} \right\} \tag{8}$$

where $q_n = \{u, v, w, \Phi, \theta, \psi\}^T$, $Q_n = \{F_x, F_y, F_z, M_x, M_y, M_z\}^T$ are the column vectors of generalized displacements and generalized forces at the n-th node; l_{mj} are (6×6) blocks of the transfer matrix; P_q, $P_Q(6xl)$ are portions of column vector of loadings components.

4.3 The Matricial Relation of Interblade Linkage

The matricial relation of interblade linkage is obtained by different technique depending on design peculiarities of linkage. For rod model of linkage the relation is obtained by TMM as it is valid for the blade. When plate model is adopted a technique based on FEM is used. This technique allows to take into account different contact conditions of shrouds and their intricate shape [4].

4.4 The Blade-Shroud Assembly Analysis

The set of resolving equations is formed by finite element method procedure which allows to consider any kind of interblade connections used in turbomachinery. The unknowns are displacement amplitudes of superelements. The superelements represent parts of blades and parts of interblade linkage between their joints. The equation system takes the form:

$$K(\Omega)q_\Sigma = P \qquad (9)$$

where $K(\Omega)$ - dynamic stiffness matrix of system; P - column vector of system loadings; $q_\Sigma = \{q_1, q_2, ..., q_N\}^T$ - column vector of displacements at nodes of the superelements.

To provide for calculations of blading with a large number of blades, the profile storage technique is used. For circular linkage of blades assembly the number of stored values is 180N-144 and for blade packet it is 108N-72, therefore the calculations may be performed even using personal computers.

Here, it should be noted next that:

1) Matrix computation (7) (8) is performed for blades with different characteristics only. Further tedious computations of definite integrals for transfer matrix components are made for the standard blade only. Section characteristics of other blades are related with characteristics of standard blade: $A = \mu^2 A^0$, $J_{xi} = \mu^4 J_\xi^0$; $J_p = \mu^4 J_p^0$ etc. Therefore the integrals of perturbed blades are calculated by multiplying integrals by perturbation parameters.

2) The forces of fluid damping and blade interaction through flow are expressed by displacements of the blades considered and of a few of their neighbors. To calculate the blade transfer matrix it is necessary to know the relationship for displacements of these blades. An iterative calculation process is used to perform these calculations. First, the amplitudes of the blade system are calculated by taking into account the aerodynamic forces which only result from the vibrations of the blade considered. The information obtained about the amplitude relationship is used to calculate the vibration by taking into account those aerodynamic forces which result from the displacements of the considered and of the neighboring blades. The iterative process converges very quickly - one or two iterations suffice as a rule.

3) Matrix of the system has a symmetrical form, but it has complex and non-symmetric matricial blocks. The lack of matricial symmetry and complex number values take care of blade material damping and phase shift in the forcing loads.

114

5. Results and Conclusions

The above presented numerical approach has been implemented in FOR-
TRAN software. The numerical simulation was concerned with forced vibrations of perturbed blade packets and optimal perturbation parameters were
looked for. By way of an example a 5 blade packet was considered. The
blades were interlinked by a shroud, load distributions along each blade was
assumed to be uniform and the damping (or dissipation) coefficients of the
blade material was 0.03. Amplitude-frequency characteristics of maximum
stresses are shown in Fig.2, Figs.2a and 2c respectively show the response
characteristic of a tuned (unperturbed) standard blade packet when excited
by harmonic components for number $k = 6$ and 38. There are five natural
frequencies in the frequency range considered, such that one is the frequency
of "in-phase" vibration and the other four are closely spaced frequencies of
the so-called "in-packet" vibration. In order to compute optimal sets of perturbation parameters a few direct search methods were assessed such as the
Hooke-Jeeves, the Powell and the coordinate descent method. Powell's method proved to be most effective for the problems considered. This method
enables computation of optimal solutions in 40 to 100 iterative steps, the
accuracy obtained being 0.1% which is satisfactory enough. Table 1 shows
the calculated sets of shape perturbation parameters $\{\mu\}_{max}$ and $\{\mu\}_{min}$ for
which σ_{max} attains a maximum and a minimum respectively within the admissible sets. σ_{max} values of the unperturbed blade packet are also given for
comparison. The admissible values of perturbation parameters range from
$0.95, \ldots, 1.05$ and σ_{max} is found within the frequency range of 0 to 2 kHz.
Response characteristics for the calculated sets are shown in Fig.(2b) ($k = 6$)
and Fig.(2d) ($k = 38$). Solid lines denote the sets $\{\mu_j\}_{max}$ and dashed lines
are for $\{\mu_j\}_{min}$.

In order to demonstrate the stress response to parameter perturbation maps
of equal levels of σ_{max} have been drawn for selected perturbation parameters.
The maps shown in Fig. 3a and 3b are for $k = 6$ and $k = 38$ respectively.
These parameters are taken for two blades of a packet. Subscripts of μ give
the blade numbers. Values on the lines are those of $10^{-1} \times \sigma_{max}$ in MPa. The
figures show the highly complex multi-extremal (or multimodal) character of
the dependence of σ_{max} on $\{\mu_j\}$. A change of the perturbation parameters
within 5 to 8% may produce a change of the stress value by a factor of two.

The above simulation technique developed and the implemented software
may serve to solve a number of important problems of turbomachinery engineering such as

- calculating the stresses which are likely to occur when the blade parameters are perturbed within manufacturing tolerances,
- determining the response of stress amplitudes in forced vibrations to
  perturbations of geometry,

- to estimate probable discrepancies of experimental and analytical results of investigations which may be due to some imprecision in describing the perturbation of the impeller blade parameters in such computations,
- to help such assembly techniques of impeller blades as to provide minimum stress levels, in particular when the manufacturing provides for control of geometry or of natural frequencies of all impeller blades either alone or in the assembly.

Table 1. The calculated sets of most favorable and most adverse perturbation parameters and the corresponding maximum stress levels.

blades numbers	$k = 6$			$k = 38$		
	$\{\mu_j\}_{max}$	$\{\mu_j\}_o$	$\{\mu_j\}_{min}$	$\{\mu_j\}_{max}$	$\{\mu_j\}_o$	$\{\mu_j\}_{min}$
1	0,959	1,0	1,039	1,005	1,0	1,050
2	0,961	1,0	1,019	0,964	1,0	0,992
3	1,039	1,0	0,976	0,956	1,0	1,003
4	1,025	1,0	0,981	0,989	1,0	1,002
5	0,995	1,0	1,050	0,999	1,0	0,997
σ_{max}, MPA	824	575	355	2116	1316	975

References

1. G. V. Reklaitis, A. Ravindran, K. M. Ragsdell "Engineering optimization methods and applications" Wiley - Interscience Publication, 1983.

2. B. F. Shorr "Bending - torsional vibration of twisted compressor blades" Strength and Dynamics of Aircraft Engines, Moscow: Machinostroenie, 1964, N1, 15-27 pp. (in Russian).

3. D. N. Gorelov, V. B. Kurzin, V. E. Saren "Album of unsteady aerodynamic characteristics of profile lattices" Novosibirsk: Nauka, 1974 (in Russian).

4. E. P. Petrov "Multifreedom finite elements models of shroud and condensation technique for vibration analysis of blades - shroud - disk assembly" Vibrational Strength and Reliability of Engines and Aircraft Structures, Samara, 1990, 121-130 pp. (in Russian).

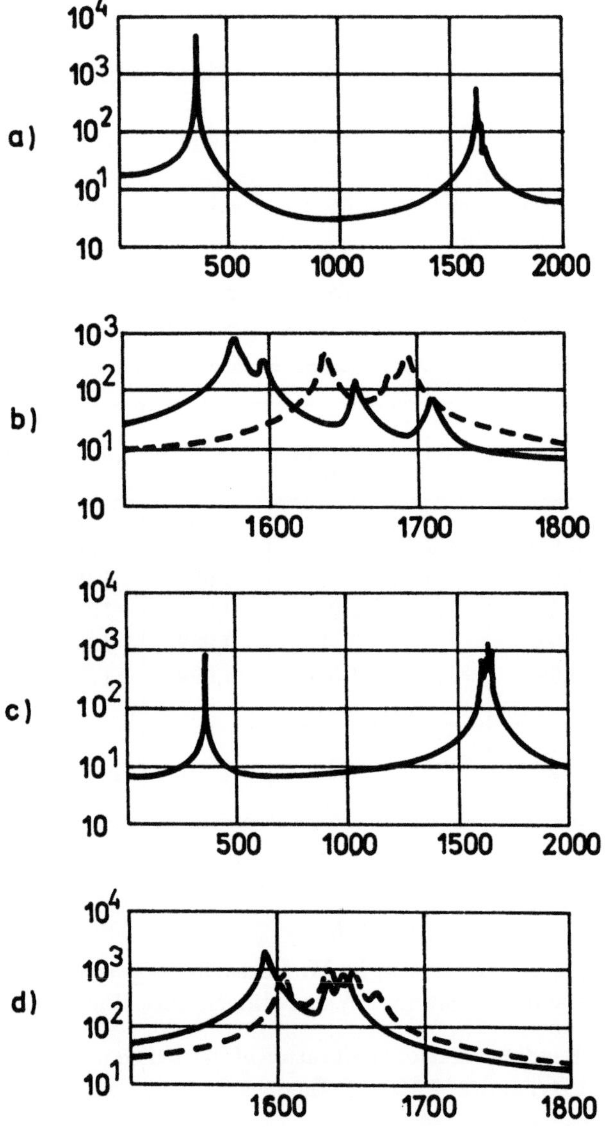

Fig. 2. Amplitude-frequency characteristics of maximum stresses: (a,c); for an unperturbed blade packet and (b,d) for a perturbed packet where perturbation parameters have been calculated (a,b) is for 6-th harmonic mode of excitation (c,d) for the 38-th harmonic mode.

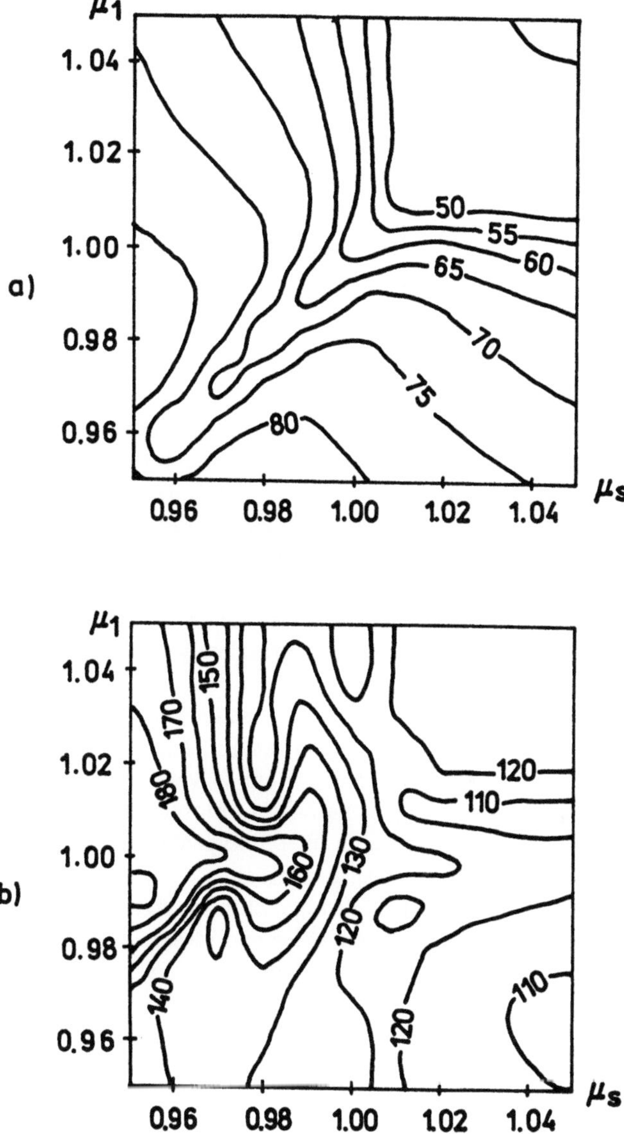

Fig. 3. The lines of equal maximum stresses v. perturbation parameters for two blades of a packet; (a) is for 6-th harmonic mode of excitation, (b) for the 38-th harmonic mode.

New Discretized Optimality Criteria Method
State of the Art

G.I.N. Rozvany and M. Zhou, FB 10, Essen University, D-45117 Essen, Germany

Abstract. It was shown recently that the optimization capability for sizing structural systems can be increased dramatically by the introduction of new optimality criteria methods which achieve the correct optimal solution while treating active stress constraints at the element level. This means that their optimization capability is only limited by the number of global (e.g. displacement or natural frequency) constraints. Since the latter is usually small for typical structural systems, these new techniques increase our optimization capability by several orders of magnitude, whilst reducing drastically both CPU time and storage requirements.

The versatility of the new optimality criteria methods is demonstrated by considering a variety of problems in sizing optimization. The proposed methods become highly economical if the considered system is very large, with many thousands of variables and active stress constraints. Moreover, the considered methods are found to be almost unavoidable in topological optimization, which will be discussed in a separate lecture at this meeting.

The paper also reviews a two-phase DCOC algorithm for structures with complex cross-sections, having several design variables and stress constraints per element. This algorithm solves the relevant equations alternately at the element level and system level.

Finally, the proposed algorithm is compared with an approach by Gutkowski, Bauer, Dems and Iwanow.

Keywords. Structural optimization, optimality criteria, Kuhn-Tucker conditions, stress constraints, displacement constraints, Lagrange multipliers, dual methods, approximations, finite elements.

1 Introduction: Methods of Structural Optimization

In this section, we discuss the advantages and disadvantages of optimality criteria methods.

Methods of structural optimization fall into one of three broad categories, namely

- optimality criteria methods;

- mathematical programming methods using gradients (sensitivities); and

- random search methods.

In *optimality criteria methods*, all necessary (and sometimes sufficient) conditions of optimality are generated for a given problem. The corresponding set of equations are then solved, partly explicitly (if possible) and partly iteratively.

In *gradient-based mathematical programming methods*, the gradients of objective (cost) function and constraint gradients are calculated for a given design and then the next design is generated in a systematic way, by trying to maximize the reduction in cost in the feasible domain (or to minimize the computational effort to find an optimum). The above procedure is repeated until the change in cost or design parameters becomes small.

Random search techniques (e.g. genetic algorithms) are based on some randomized procedure which generates systematically a large number of designs, out of which the best one is selected.

A method is termed *robust*, if it can be used "blindly" or as a "black box" for a broad range of problems, without having to understand at depth the particular mathematical structure of the optimal solutions involved. Clearly, search methods are, in general, the most robust and optimality criteria methods the least robust.

A method is called *efficient* if it has a higher optimization capability in terms of number of variables or number of active constraints, or alternatively, if it uses less computer time and/or storage space than other methods. The relative efficiency of optimality criteria methods is usually the highest and that of search methods is, in general, the lowest. However, search methods may be unavoidable if the number of local minima is very high.

Gradient-based *mathematical programming* methods may also be divided into two subclasses, i.e.

- *primal methods*, in which the variable space contains the design variables; and

- *dual methods*, in which the variables are the Lagrange multipliers of the constraints.

Clearly, primal methods are uneconomical if the number of variables is large and dual methods are inefficient if the number of active constraints is too high. This means, however, that for problems with many variables and many active constraints, both primal and dual methods are inadvisable. This situation may appear hopeless; fortunately, however, most active behavioural constraints for structural systems are *local* (stress or side) *constraints* and these can often be eliminated in modern optimality criteria methods (COC/DCOC) through explicit calculations at the element level. This means that, in effect, only *global* constraints contribute to the size of the dual space, and hence the above methods become several orders of magnitude more efficient for very large structural systems than traditional techniques.

Another way of avoiding stress constraints in the dual space is to use a *fully stressed design* (FSD) for stress controlled elements and to treat other constraints properly in the dual space. This method, however, is known to result, in general, in non-optimal solutions.

We may add that the above classifications are somewhat oversimplified. It can be shown that certain dual programming methods and optimality criteria methods are, in fact, identical. Moreover, optimality criteria can be expressed through sensitivities and hence the distinction between programming and optimality criteria methods is far from being sharp.

The newest optimality criteria method (DCOC) can also be regarded as a modified version of the older discretized optimality criteria (DOC) method combined with fully stressed design (DOC-FSD). The difference is that DCOC is based on rigorous fulfilment of the Kuhn-Tucker conditions, whereas DOC-FSD is an intuitive method. Whilst both methods use the same redesign formulae, in DCOC the virtual load systems, corresponding to displacement constraints, are modified through the application of prestrains, in order to fulfil the above conditions. The details of this operation are explained in Section 3.

2 The Origins of Optimality Criteria Methods

The earliest optimality criteria methods were somewhat heuristic. These *intuitive optimality* criteria methods included the already mentioned *fully stressed design* (FSD), as well as *uniform energy dissipation* (UED), and *unifom mutual energy dissipation* (UMED). All the above methods give the correct optimal solution for a limited class of problems: FSD for statically determinate structures, UED for a compliance (total work) constraint and UMED for a displacement constraint, but all three are restricted to a single load condition.

Rigorous optimality criteria were derived for *discretized systems* by Berke and others (e.g. Berke 1970; Venkayya, Khot and Berke 1973). These optimality conditions were then modified, on a somewhat intuitive basis, into very efficient redesign formulae. This group of methods will be termed here DOC (discretized optimality criteria) methods.

Surprisingly, entirely independently from the above development but almost at the same time, a research school around Prager derived optimality criteria for continua (e.g. Prager and Shield 1967; Prager and Taylor 1968; for eigenvalue problems, see e.g. Olhoff 1981). The above conditions were derived usually for a single criterion and often for plastic design. However, they were gradually extended to general elastic systems with simultaneous stress and displacement constraints (e.g. Rozvany 1989).

One of the obvious differences between the above developments was the fact that one (DOC) used a discretized formulation in a finite dimensional vector space and the other one (COC) a continuum formulation in an infinite dimensional design space. Correspondingly, the branch of mathematics involved is differential calculus in one case (DOC) and calculus of variations in the other (COC). A recently found, more intrinsic difference is that, whereas in DOC the only variables are the cross-sectional dimensions (design variables), in COC an extended design space includes also the real and virtual internal forces. Another improvement is the introduction of an "adjoint structure" in COC, which is used for giving the optimality conditions some physical meaning. The external loads on the adjoint structure depend on active displacement constraints and the fictitious prestrains in members on active stress constraints (see Section 3).

The COC algorithm was applied iteratively to large discretized systems (Rozvany and Zhou *et al.* 1989, 1990; Rozvany and Zhou 1991b; Zhou and Rozvany 1993a), by first deriving the optimality criteria for continua and then discretizing the results for FE methods. This transformation introduced extra work and the results were only valid for a continuous variation of the design variables along the members or elements involved. For these reasons, it represented a considerable improvement when Zhou reformulated the new optimality criterion method directly in terms of matrix methods of the finite element approach (Zhou and Rozvany 1992, 1993b). The new *discretized continuum type optimality criteria method* has been termed DCOC for historical reasons. As will be seen in Section 3, DCOC was first derived using the flexibility formulation of structural analysis and the fundamental features of COC, employing a stiffness method in the implementation only. This has had the drawback that it was difficult to understand by researchers familiar with traditional optimality criteria (DOC) methods. Moreover, the implementation was cumbersome for more advanced elements whose flexibility relationships are not readily available in the modern FE technology. These difficulties have been removed by recent work of Zhou and Haftka (1994) who showed that DCOC can, in fact, be derived from the traditional DOC formulation. In the above paper a new, "derivative-based" DCOC formulation is also presented, which avoids the flexibility formulation completely, as well as the concept of adjoint structure, replacing the latter with sensitivity analysis of certain functions. This way, implementation of DCOC becomes very convenient.

It should be mentioned in any text on modern optimality criteria methods that another important group of dual methods was developed by a Polish research team (Bauer, Gutkowski and Iwanow 1981; Gutkowski, Bauer and Iwanow 1985; Dems and Gutkowski 1993). One of the differences between COC/DCOC and the above methods is that in the latter the formulation includes compatibility conditions, whilst in the former they turn out to be optimality criteria.

In the next section, however, COC/DCOC will be outlined on the basis of their original formulation, because this will have some conceptual advantages in discussing topology optimization in later sections.

3 The COC/DCOC Algorithm: a Brief Review

3.1 The Role of the "Adjoint Structure"

As mentioned in Section 2, a fundamental conceptual feature of the new optimality criteria methods is the adjoint structure, which is a fictitious system having

122

the same analysis equations as the real structure but different loads, prestrains and possibly support settlements. The analysis of the adjoint structure replaces the usual sensitivity analysis of other methods, although it can be shown in some cases to reduce to conventional sensitivity analysis by the adjoint variable method. The implementation of the adjoint method is rather efficient, because the stiffness matrix is the same for the real and adjoint structures. As pointed out earlier, this concept can also be avoided to suit existing computational technology.

In this paper, we shall not include the derivation or the general formulae of COC and DCOC, because they are summarized in a comprehensive form elsewhere (Zhou and Rozvany 1992, 1993a; for a detailed summary see Rozvany and Zhou 1994). For *didactic reasons*, we are only considering herein *truss structures*, and illustrations through very simple examples. The DCOC method is, however, available for most static and some dynamic problems, including a number of refinements (see Rozvany and Zhou 1994 or Section 6 herein).

To explain the concept of the adjoint structure, consider the elementary problem in Fig. 1a, in which a three-bar truss is subject to the horizontal force P_1 and the vertical force $P_2 = kP$ and the weight of the truss is minimized subject to constraints on the stresses in compression and tension (σ_C and σ_T), and on the vertical deflection ($U_2 \leq t$). The horizontal displacement (U_1) is unconstrained. One of the shortcomings of this problem is the fact that a vertical displacement constraint is equivalent to a stress constraint for the vertical bar "2". This means that, in general, depending on the limiting values σ_T and t in the above problem, only one of the above constraints can be active. However, this restricting feature does not disturb the illustration of the DCOC method.

The *adjoint structure* for the above problem is shown in Fig. 1b. The loading on the adjoint structure is the Lagrange multiplier ν for the displacement constraint, acting at the location and in the direction of that constraint. The initial displacements (caused by fictitious prestrains) for the adjoint structure are $\overline{\delta}^{*(1)}$, $\overline{\delta}^{*(2)}$ and $\overline{\delta}^{*(3)}$, which are non-zero only if a stress constraint for a bar is active. For trusses, the above initial displacement is given by (Rozvany and Zhou 1994)

$$\overline{\delta}^{*e} = \frac{L^e}{\sigma^e}\left[\rho^e - \frac{f^e\overline{f}^e}{(x^e)^2 E^e}\right]\mathrm{sgn}\,(f^e),\tag{1}$$

where $L^e, \sigma^e, \rho^e, E^e$ and x^e are the bar length, permissible stress, specific weight, Young's modulus and cross-sectional area for the truss element e, whilst f^e and $\overline{f}^e$ are the corresponding bar forces in the real and adjoint trusses.

It can be seen that the adjoint structure is equivalent to the usual virtual load system with a scaling factor ν (in DOC or adjoint variable method of sensitivity analysis), and prestrains added for active stress constraints.

3.2 Iterative Procedure for COC/DCOC

The basic steps in the new optimality criteria methods are as follows.

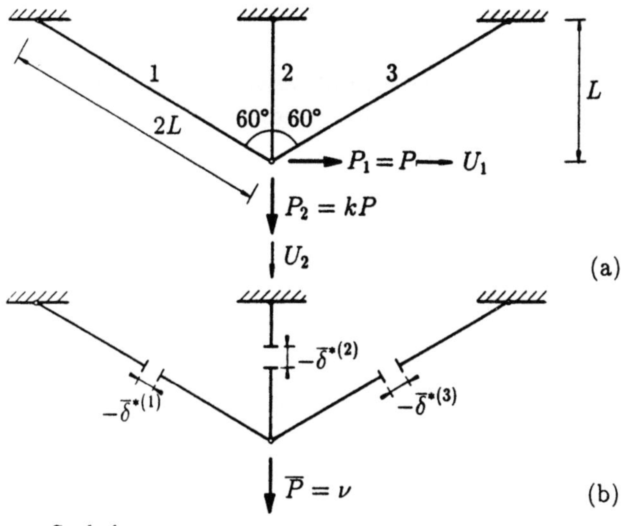

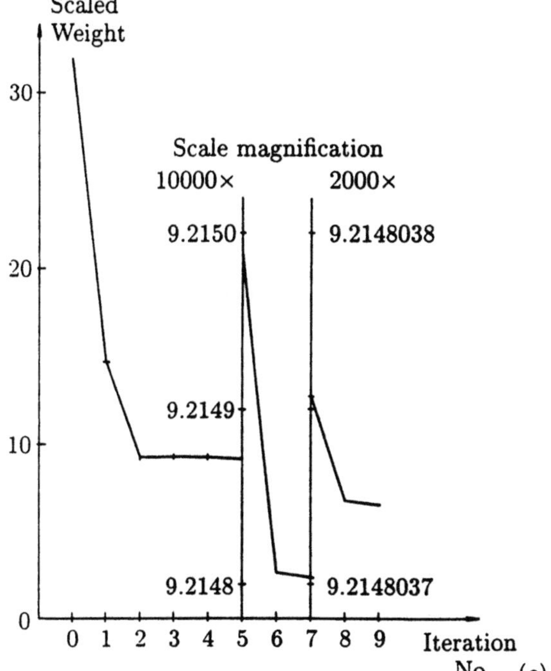

Fig. 1. Elementary example.

Details of the latter two operations are given below.

- Analysis of the real and adjoint structures.
- Updating cross-sectional parameters.
- Updating Lagrange multiplier(s) for global constraint(s).

124

3.3 Updating Cross-Sections

The redesign formula used depends on the type of active constraints for a given element. In numerical procedures, it is usually necessary to impose at least a lower limit on the cross-sectional parameter, e.g.

$$x^2 \leq x_{\downarrow}^e , \tag{2}$$

where $x_{\downarrow}^e$ is a given minimum cross-sectional area for a bar e. Such restrictions are termed *side constraints*.

For trusses with stress and side constraints and a displacement constraint (e.g. Fig. 1), we have the following redesign formulae:

$$(x^e)^2 = \max \left\{ (x_d^e)^2, \ (x_\sigma^e)^2, \ (x_{\downarrow}^e)^2 \right\} , \tag{3}$$

with

$$(x_d^e)^2 = \frac{f^e \overline{f}^e}{E^e \rho^e} , \tag{4}$$

$$(x_\sigma^e) = \frac{|f^e|}{\sigma^e} . \tag{5}$$

Clearly, (5) determines the cross-sections controlled by the stress constraint. The redesign relation (4) refers to truss members controlled by the displacement constraint. In general, elements of a truss fall into one of the three element sets, namely [see (3) above]

R_d : elements controlled by the displacement constraints,

R_σ : elements controlled by a stress constraint,

R_s : elements controlled by a side constraint.

3.4 Updating Lagrange Multipliers

Since $\overline{f}^e$ in (4) depends on the Lagrange multiplier ν (see Fig. 1b), the updated value of x_d^e also depends on it, whereas $x_{\downarrow}^e$ and x_s^e are independent of ν. This means that by any change of ν only the elements in R_d are affected. The new value of ν can therefore be calculated by satisfying the *displacement constraint* as an equality using a work equation, substituting the cross-sectional areas from (4) and expressing ν from the above relation:

$$\sqrt{\nu} = \frac{\sum\limits_{e \in R_d} L^e \sqrt{\rho^e f^e \hat{f}^e / E^e}}{t - \sum\limits_{e \notin R_d} (L^e f^e \hat{f}^e / E^e x^e)} , \tag{6}$$

in which $\hat{f}^e = \overline{f}^e / \nu$, with the value of ν taken from the prior iteration.

Using the above procedure for the elementary problem in Fig. 1a, with $L^e = E^e = \rho^e = 1$ and $\sigma_t^e = \sigma_C^e = 1.3$ for *all* elements as well as $P = P_1 = 1$, $k = P_2 = 8$ and $t = 1$, the iteration history is shown in Fig. 1c, in which an almost uniform convergence is achieved in spite of scale magnifications by 10000 and 2000. The optimal weight w_{opt} after 9 iterations was found to be 9.2148037239 and the cross-sectional areas $x^{(1)} = 0.882077$, $x^{(2)} = 7.430649$ and $x^{(3)} = 0.01$. In iteration 9, the weight value changed only from 9.2148037242 to the above value, satisfying the very stringent convergence criterion

$$\frac{|W_{\text{new}} - W_{\text{old}}|}{W_{\text{new}}} \leq T, \tag{7}$$

with $T = 10^{-10}$. For practical applications such high accuracy is not necessary. The above iteration history (Fig. 1c) shows that after two iterations the (scaled) weight differs by only one per cent from the optimal weight.

3.5 Some Philosophical Remarks about Conceptual Clarity vs. Computational Expediency

The above dilemma will be illustrated through the formula for calculating the adjoint initial displacement in (1). Since for stress-controlled truss elements we have $f^e/x^e = \sigma^e \text{sgn}\,(f^e)$, the relation in (1) reduces to

$$\overline{\delta}^{*e} = L^e \left[\frac{\rho^e \text{sgn}\,(f^e)}{\sigma^e} - \frac{\overline{f}^e}{x^e E^e} \right]. \tag{8}$$

However, the external load on the adjoint truss causes an elastic member elongation of $L^e \overline{f}^e / x^e E^e$. Adding the prestrain to the strain due to external load, we obtain the total adjoint strain

$$\overline{\varepsilon}^e = \frac{\rho^e \text{sgn}\,(f^e)}{\sigma^e}, \tag{9}$$

which is simply *the ratio of the specific weight and the permissible stress*, with the sign of the real force in the considered truss element. Whilst the result in (9) is conceptually very valuable due to its simplicity, *in an iterative procedure* various quantities in (8) originate from different computational steps. This means that *for intermediate iterative values* of $\overline{f}^e$ (involving ν) and x^e, $\overline{\varepsilon}^e$ does not reduce to (9). Moreover, it is difficult to model the requirement that the total adjoint strain in stress-controlled members is independent of the adjoint force in the same member. Surprisingly, relation (1) is therefore more efficient computationally than (9).

Naturally, in all practical computations the most expedient formulae should be used, even when the latter completely obscures conceptually fundamental features of an optimal solution. In order to be able to check the correctness of discretized numerical solutions, or with a view to obtaining exact, explicit solutions (for example, in topology optimization), researchers should also be thoroughly familiar with the original, exact relations leading to a computational procedure.

4 Iterative COC/DCOC: Approximations Used and Their Effect on Computational Stability

Update formulae in the above iterative procedures involve a certain approximation, because *the effect of the cross-sectional areas on the internal forces is temporarily ignored.* The same applies to more complicated DCOC procedures for several load conditions and multiple displacement constraints, where the Lagrange multipliers for displacement constraints are calculated in a subiteration (Zhou and Rozvany 1993b). All the above formulae would be exact if the above dependence did not exist, that is, in the case of *statically determinate structures.* It follows from the above approximation concept that DCOC converges very satisfactorily if *the distribution of the internal forces does not depend significantly on the cross-sectional areas.* Convergence problems arise, and further refinements become necessary, if the problem is ill-conditioned in the sense that *the internal forces are highly dependent on the cross-sectional areas.* To avoid computational instability in the latter case, a procedure for automatic adjustments of the move limits can be adopted. A better solution of this problem is to introduce an improved approximation for the updating procedure.

It is further to be remarked that the redesign formula (4) is based on first order sensitivities but the relation in (5) is not related to derivatives. The latter, therefore, results in a slower convergence. In *DOC and dual methods*, stress constraints are usually replaced by equivalent displacement constraints, and hence first order redesign formulae are involved. This way the number of iterations is reduced, but the total computer time is increased because the calculation of the corresponding Lagrangians, which are coupled at the system level, becomes much more expensive.

5 Iterative COC/DCOC: a Review of Test Examples

A number of test examples based on DCOC were presented in recent publications (Zhou and Rozvany 1992, 1993b; Rozvany and Zhou 1994). Since for trusses and beams the COC and DCOC procedures are somewhat similar, earlier test examples using an iterative COC procedure are also relevant (Rozvany, Zhou *et al.* 1989, 1990). The above examples included

- a clamped beam with stress constraints and a displacement constraint;

- ten-bar truss examples with (i) stress constraints and one displacement constraint, (ii) stress constraints and two displacement constraints, (iii) only stress constraints with different permissible stress values, and (iv) two displacement constraints and a variable point load;

127

- a ten-storey, three-bay frame with stress constraints and a horizontal displacement constraint;

- the standard 25-bar truss;

- the standard 72-bar truss;

- the standard 200-bar truss problem and a modified version of it.

Out of the above test problems, the 25-bar and 72-bar trusses will be reviewed here in detail.

5.1 25-bar truss

The standard 25-bar truss is shown in Fig. 2. The material properties are: E $= 10^7$ psi, $\gamma = 0.1$, lbs/in^3 and the lower limit on the design variables $y^L = 0.01$ in^2.

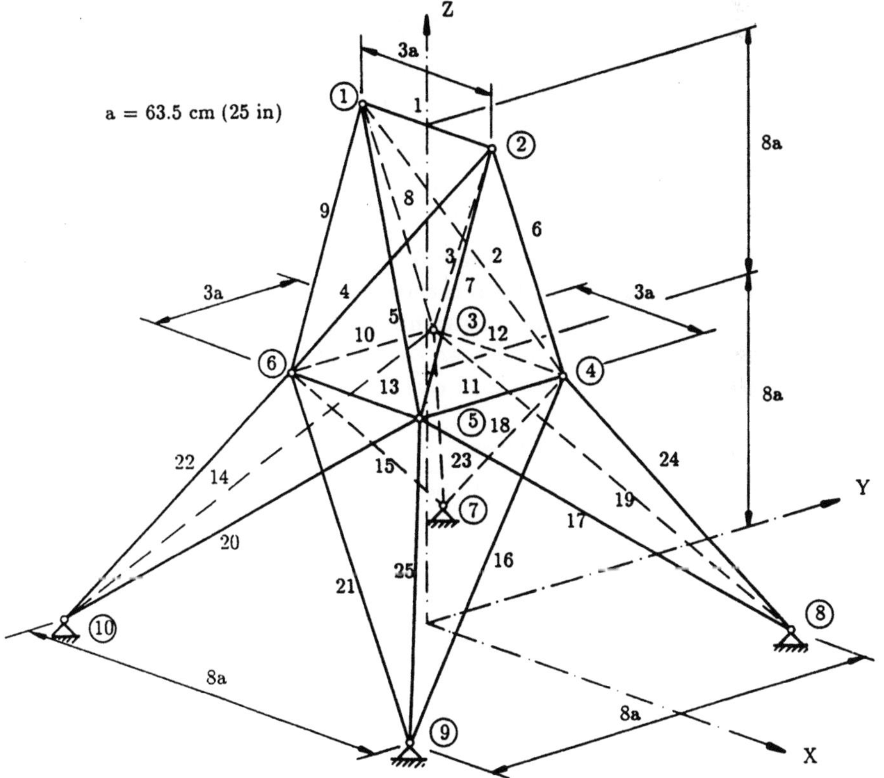

Fig. 2. 25-bar truss.

The allowable displacements for all nodes are ± 0.35 in in the X, Y and Z directions. The loading and allowable stresses are given, respectively, in Tables

1 and 2. The 25 members are linked to 8 variables. For additional details see Fleury and Schmit (1980).

Table 1. Allowable stresses for 25-bar truss (psi)

Member	Tension	Compression
1	40000	-35092
2-5	40000	-11590
6-9	40000	-17305
10, 11	40000	-35092
12, 13	40000	-35092
14-17	40000	-6759
18-21	40000	-6759
22-25	40000	-11082

Table 2. Nodal load components (lbs) for the 25-bar truss

Load case	Node	x	y	z
1	1	1000	10000	-5000
	2	0	10000	-5000
	3	500	0	0
	6	500	0	0
2	1	0	20000	-5000
	2	0	-20000	-5000

The final solutions are given in Table 3. The cross-sectional distribution of the final design for DCOC and DUAL is shown in Fig. 3. For all three methods, the final design has the following constraints active: the displacements at node 1 and 2 in the Y direction for both load cases and the compression stresses at elements 19 and 20 for the second load case.

Table 3. Results for the 25-bar truss

Linked variables	Member	Cross-sectional area (in^2)		
		DCOC	DUAL	DOC-FSD
1	1	0.0100	0.0100	0.0100
2	2-5	1.9870	1.9870	1.9747
3	6-9	2.9935	2.9935	3.0120
4	10, 11	0.0100	0.0100	0.0100
5	12, 13	0.0100	0.0100	0.0100
6	14-17	0.6840	0.6840	0.6848
7	18-21	1.6769 (s)	1.6769 (s)	1.6789 (s)
8	22-25	2.6621	2.6621	2.6557
No. of analyses		41	15	38
Weight (lb)		545.162710	545.162710	545.167087
CPU time (sec.)	Opt.	0.22	0.10	0.16
	Total	1.05	0.38	0.94

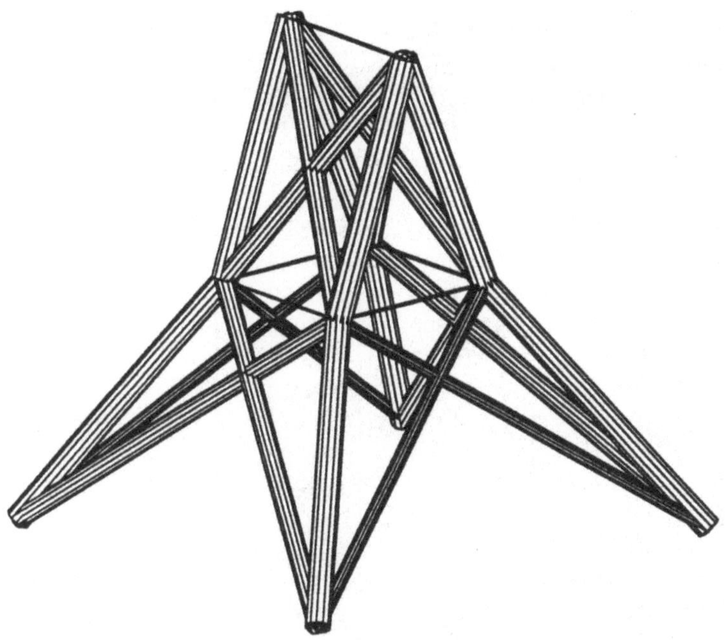

Fig. 3. Distribution of the final cross-sections for the 25-bar truss, $x_{\min} = 0.01$, $x_{\max} = 2.993$.

5.2 72-bar truss

The considered problem is shown in Fig. 4. The material properties are: $E = 10^7$ psi, $\gamma = 0.1$, lbs/in^3, $\sigma_a = 25000$ psi and the lower limit on the design variables $y^L = 0.1$ in^2.

The loading is shown in Table 4. The allowable displacements for all nodes in x and y directions are ± 0.25 in. 72 members are linked to 16 variables. For additional detials see Fleury and Schmit (1980).

Table 4. Nodal load components (lbs) for the 72-bar truss

Load case	Node	x	y	z
1	1	5000	5000	-5000
2	1	0	0	-5000
	2	0	0	-5000
	3	0	0	-5000
	4	0	0	-5000

The final solutions are given in Table 5. The cross-sectional distribution of the final design for DCOC and DUAL is shown in Fig. 5. For all three methods, the final design has the following constraints active: displacements at node 1 in both X and Y directions for load case 1, and the stress constraints at elements 1-4 for load case 2 (four dependent constraints).

Table 5. Results for the 72-bar truss

Linked variables	Member	Cross-sectional area (in^2)		
		DCOC	DUAL	DOC-FSD
1	1-4	0.15646 (s)	0.15646 (s)	0.15712 (s)
2	5-12	0.54560	0.54560	0.53559
3	13-16	0.41038	0.41038	0.40960
4	17, 18	0.56975	0.56975	0.56927
5	19-22	0.52368	0.52368	0.50666
6	23-30	0.51710	0.51710	0.51997
7	31-34	0.10000	0.10000	0.10000
8	35, 36	0.10000	0.10000	0.10000
9	37-40	1.26835	1.26835	1.28016
10	41-48	0.51165	0.51165	0.51478
11	49-52	0.10000	0.10000	0.10000
12	53, 54	0.10000	0.10000	0.10000
13	55-58	1.88619	1.88619	1.89726
14	59-66	0.51231	0.51231	0.51576
15	67-70	0.10000	0.10000	0.10000
16	71, 72	0.10000	0.10000	0.10000
No. of analyses		10	10	18
Weight (lbs)		379.614802	379.614802	379.657382
CPU time (sec.)	Opt.	0.06	0.06	0.18
	Total	1.21	0.98	1.98

6 Recent Extensions of the DCOC Method

Current extensions of DCOC include (e.g. Rozvany and Zhou 1994)

- stress constraints,

- several load conditions,

- several displacement constraints per load condition,

- elastic supports,

- given support settlements,

- contact problems,

- allowance for the cost of reactions,

- variable loads,

- variable prestrains,

- variable support settlements,

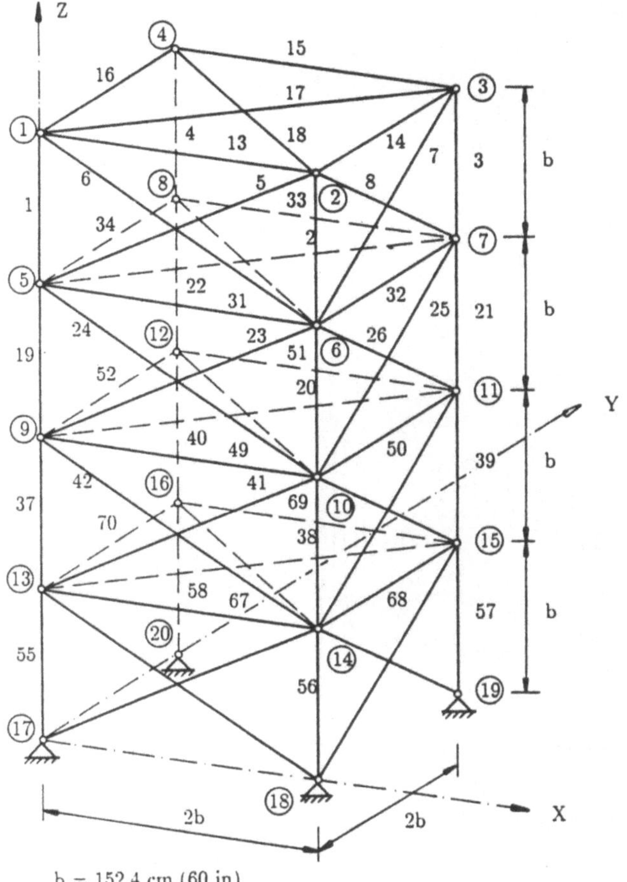

Fig. 4. 72-bar truss.

- allowance for the cost of variable loads,
- allowance for the cost of variable prestrains,
- allowance for the cost of variable support settlements,
- allowance for selfweight,
- allowance for member buckling,
- system stability constraints,
- natural frequency constraints,
- material nonlinearity,
- structures with complex cross-sections.

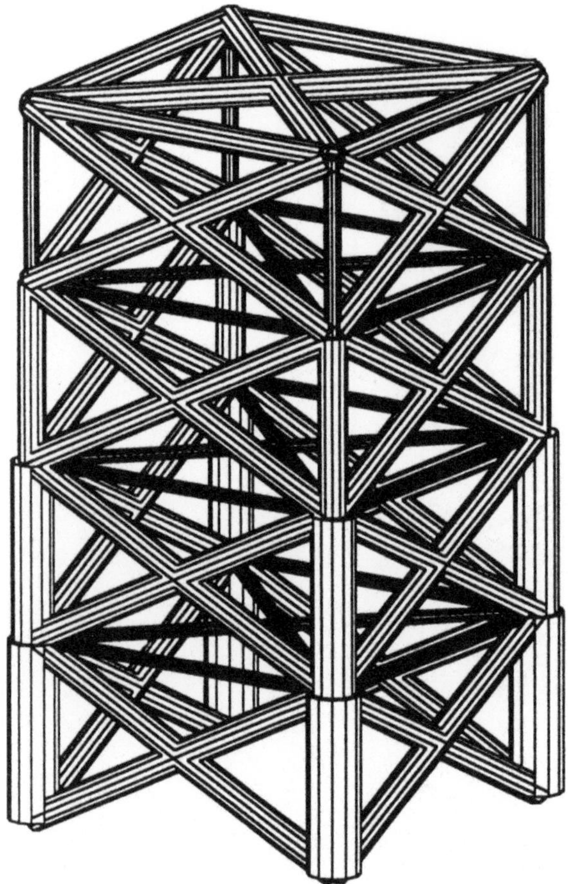

Fig. 5. Distribution of the final cross-sections for the 72-bar truss, $x_{\min} = 0.1$, $x_{\max} = 1.886$.

Extensions to geometrical nonlinearity and to probabilistic design (which is subject of this meeting) are being developed currently.

As regards the type of structures used in test examples, DCOC has been applied to trusses, beams, frames, grillages, disks and plates. Extensions to shells and other structures are straightforward.

The last extension is reviewed briefly below.

6.1 DCOC Algorithm for Sizing Structures with Complex Cross-Sections

For problems involving several sizing variables for each element (e.g. typical frame elements used in practice), the element level variables, i.e. cross-sectional dimensions and Lagrangian multipliers associated with stress and side constraints, are governed by several simultaneous nonlinear equations. Moreover, these element level equations are also influenced by Lagrangian multipliers associated with displacement constraints. The solution of the governing equations in the DCOC method becomes, therefore, more complicated. Moreover, a two-phase iterative algorithm, which solves alternately the equations at the element level and system level, can be used. For both phases the Newton iterative algorithm is employed. For given values of Lagrangian multipliers associated with displacement constraints, *the element level solver* solves the same number of separable sets of equations as the number of elements, to obtain the cross-sectional dimensions and Lagrangian multipliers associated with stress and side constraints, and also provides the sensitivities of design variables with respect to multipliers associated with displacement constraints. The latter is needed in the *system level solver* to generate iteratively the Lagrangian multipliers associated with active displacement constraints on the basis of equations governing those constraints. After each analysis, the above two phases have to be performed alternately until the overall iterations converge. Then a new analysis is performed and the above procedure is applied again. This iterative algorithm is repeated until an optimal solution is achieved. It is important to note that the high efficiency of the DCOC method holds for this class of problems as well since the computational effort associated with the element level operation increases only linearly with the increase of the number of design variables, and the system level dual problem involves only Lagrangian multipliers associated with active displacement constraints, the number of which is usually much smaller than the number of active stress constraints.

7 Concluding Remarks

A brief state-of-the-art review of the new optimality criteria methods COC/DCOC was given in this lecture. Applications of these methods in topology optimization will be discussed in the second lecture.

8 References

Bauer, J.; Gutkowski, W.; Iwanow, Z. 1981: A discrete method for lattice structures optimization. *Engrg. Optim.* **5**, 121–128.

Berke, L. 1970: An efficient approach in the minimum weight design of deflection limited structures. *AFFDL-TM-70-FDTR*.

Dems, K.; Gutkowski, W. 1993: Optimal design of truss configuration under mulitload conditions. In: Herskovits, J. (Ed.) *Proc. Int. Conf. "Structural Optimization 93"*, pp. 29–36, Rio de Janeiro.

Fleury, C; Schmit, L.A. 1980: Structural synthesis by combining approximation concepts and dual methods. *AIAA J.* **18**, 1252-1260

Gutkowski, W.; Bauer, J.; Iwanow, Z. 1985: Discrete structural optimization. *Comp. Meth. Appl. Mech. Eng.* **51**, 71–78.

134

Olhoff, N. 1981: Optimization of columns against buckling. Optimization of transversely vibrating beams and rotating shafts. In: Haug, E.J.; Cea, J. (Eds.) *Optimization of Distributed Parameter Structures* (Proc. NATO ASI, Iowa 1980), pp. 152–176, 177–199, Sijthoff and Noordhoff, Alphen aan der Rijn.

Prager, W.; Shield, R.T. 1967: A general theory of optimal plastic design. *J. Appl. Mech.* **34**, 184–186.

Prager, W.; Taylor, J. 1968: Problems of optimal structural design. *J. Appl. Mech.* **5**, 102–106. Rozvany, G.I.N. 1989: *Structural Design via Optimality Criteria*, Kluwer, Dordrecht.

Rozvany, G.I.N.; Zhou, M. 1991b: The COC algorithm, Part I: cross-section optimization or sizing (presented 2nd World Congr. Comp. Mech., Stuttgart, 1990). *Comp. Meth. Appl. Mech. Eng.* **89**, 281–308.

Rozvany, G.I.N.; Zhou, M. 1994: Optimality Criteria Methods for Large Structural Systems. Chaper 2 in: Adeli, H. (Ed.) *Advances in Design Optimization*. Chapman and Hall, London.

Rozvany, G.I.N.; Zhou, M.; Gollub, 1990: Continuum-type optimality criteria methods for large finite element systems with a displacement constraint – Part II. *Struct. Optim.* **2**, 77–104.

Rozvany, G.I.N.; Zhou, M.; Rotthaus, M.; Gollub, W.; Spengemann, F. 1989: Continuum-type optimality criteria methods for large systems with a displacement constraint – Part I. *Struct. Optim.* **1**, 47–72.

Venkayya, V.B.; Khot, N.S.; Berke, L. 1973: Application of optimality criteria approaches to automated design of large practical structures. *AGARD-CP-123*, pp. 3.1–3.9.

Zhou, M.; Haftka, R.T. 1994: A comparison study of optimality criteria methods. *Proc. 35th AIAA/ASME/ASCE/AHS/ASCSDM Conf.* (held in South Carolina, April 1994) (to appear).

Zhou, M.; Rozvany, G.I.N. 1991: The COC algorithm, Part II: topological, geometrical and generalized shape optimization. *Comp. Meth. Appl. Eng.* **89**, 309–336.

Zhou, M.; Rozvany, G.I.N. 1992: DCOC: an optimality critria method for large systems, Part I: theory. *Struct. Optim.* **5**, 12–25.

Zhou, M.; Rozvany, G.I.N. 1993a: Iterative COC methods– Parts I and II. In: Rozvany, G.I.N. (Ed.) *Optimization of Large Structural Systems* (Proc. NATO/DFG ASI, Berchtesgaden 1991), pp. 27–75, Kluwer, Dordrecht.

Zhou, M.; Rozvany, G.I.N. 1993b: DCOC: an optimality critria method for large systems, Part II: algorithm. *Struct. Optim.* **6**, 250–262.

Discretized Method for Topology Optimization

G.I.N. Rozvany and M. Zhou, T. Birker, T. Lewiński, FB 10, Essen University, Postfach 103764, D-45117 Essen, Germany

Abstract. After discussing the aims and significance of topology optimization, layout optimization of grid-like structures is discussed in detail. In the proposed method, the ground structure (structural universe) approach is used, together with continuum-type optimality criteria. Layout optimization may be carried out using (i) analytical methods for exact, often closed-form, continuum-type solutions or (ii) numerical methods for approximate, discretized solutions. Due to the high optimization capability of modern OC methods, the latter provides simultaneous optimization of geometry and topology, with a high degree of accuracy. Finally, a new, extended optimal layout theory for multiload, multipurpose structures is outlined and applications to particular classes of problems given.

Keywords. Structural optimization, toplogy optimization, layout optimization, shape optimization, finite elements, trusses, displacement constraints, stress constraints, multiple loads, multipurpose structures, perforated plates, adjoint methods..

1 Introduction

In this section, we shall discuss the scope, aims and significance of topology optimization.

Topology of a structural system means the spatial sequence or configuration of members and joints or internal boundaries. The two main fields of application of topology optimization are layout optimization of grid-like structures and generalized shape optimization of continua or composites.

A *grid-type structure* has the basic feature that it consists of a system of intersecting members, the cross-sectional dimensions of which are small in comparison to their length, and hence the members can be idealized as one-dimensional continua. Consequences of this feature are that

- the influence of member intersections on strength, stiffness and structural weight can be neglected, and
- the specific cost (e.g. structural weight per unit area or volume) can be expressed as the sum of the costs (weights) of the members running in

various directions. Examples of grid-type structures are trusses, grillages (beam systems), shell-grids and cable nets.

Layout optimization of grid-type structures consists of three simultaneous operations, namely

- *topological optimization* involving the spatial sequence of members and joints,

- *geometrical optimization* involving the coordinates of joints, and

- *sizing*, i.e. optimization of cross-sectional dimensions.

The above concepts are explained on an example in Fig. 1, in which all three trusses have the same topology, whilst the trusses in Figs. 1b and c have the same geometry but different cross-sectional dimensions.

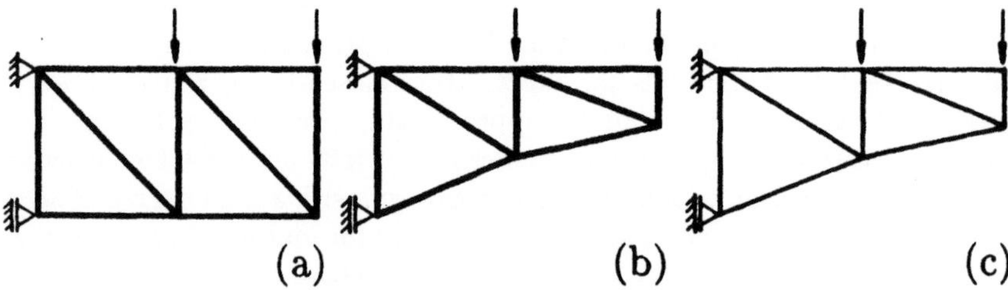

(a)   (b)   (c)

Fig. 1. Example illustrating topological, geometrical and cross-sectional properties of a grid-type structure.

Prager (Prager and Rozvany 1977b) regarded layout optimization as the *most challenging* class of problems in structural design because there exists an infinite number of possible topologies which are difficult to classify and quantify; moreover, at each point of the available space potential members may run in an infinite number of directions. At the same time, layout optimization is of *considerable practical importance*, because it results in much greater material savings than pure cross-section (sizing) optimization.

The other important field of application of topology optimization is *generalized shape optimization*, in which a simultaneous optimization of both *topology and shape of boundaries* is required for continua and *of interfaces* between materials for composites. Fig. 2a, for example, shows the initial boundary shape and topology and Fig. 2b a hypothetical optimal shape and topology for a composite plate in plane stress, where the dotted regions denote the less stiff, weaker and lighter material. For a cellular structure (here: perforated plate), the dotted regions denote cavities (or "holes"). Generalized shape optimization is discussed elsewhere (e.g. Rozvany, Zhou and Sigmund 1994).

In the next two sections, a general description of a general theory of optimal layouts is given and its application to the derivation of exact analytical solutions is presented.

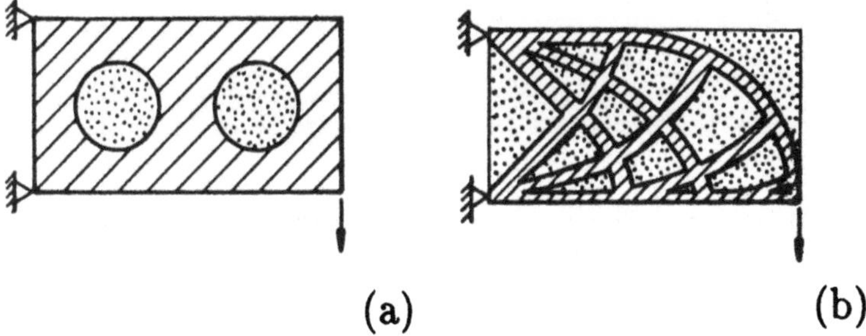

Fig. 2. Example illustrating generalized shape optimization.

2 Theory of Optimal Structural Layouts

Optimal layout theory, developed in the seventies by Prager and the first author (e.g. Prager and Rozvany 1977a and b) as a generalization of Michell's (1904) theory for trusses, was originally formulated for grid-like structures on the basis of the simplifying assumptions listed in Section 1. These simplifications were removed in a more advanced version of layout theory (e.g. Rozvany, Olhoff, Cheng and Taylor 1982; Rozvany, Olhoff, Bendsøe *et al.* 1987), which was applied to structures in which a high proportion of the available space was occupied by material. One could also say that the original, "classical" layout theory is concerned with structures having a low "volume fraction" (i.e. material volume/available volume ratio), whilst advanced layout theory deals with structures having a high volume fraction. In the latter, first the microstructure is optimized for a given ratio of the stiffnesses or forces in the principal directions and in a second operation, the optimal macroscopic distribution of microstructures is determined using methods of the layout theory.

The classical layout theory is based on *two underlying concepts*, namely

- the *structural universe* (in numerical methods: ground structure), which is the union of all potential members or elements, and

- *continuum-type optimality criteria* (COC), expressed in terms of a fictitious system termed *adjoint structure*, which were discussed in detail in the first lecture).

Since the above optimality criteria also provide adjoint strains for vanishing members, their fulfilment for the entire structural universe represents a necessary and sufficient condition of layout optimality if the problem is convex and certain additional requirements (e.g. existence) are satisfied. The above condition of convexity is fulfilled for certain so-called "self-adjoint" problems, the analytical treatment of which will be discussed in Section 3.6.

On the basis of optimal layout theory, two basic approaches have been developed using, respectively,

- *analytical methods* for deriving *closed form continuum-type solutions* representing the *exact optimal layout*, and

- *numerical, discretized iterative methods* for deriving *approximate* (but usually highly accurate) *optimal layouts.*

Fundamental differences between these two methods are listed in Table 1.

Table 1. A comparison of analytical and numerical methods based on layout theory

Computational Method	Analytical	Numerical
Structural Model	Continuum	Discretized (Finite Elements)
Procedure	Simultaneous Solution of All Equations	Iterative Solution
Structural Universe	Infinite Number of Members	Finite but Large Number of Members (Several Thousands)
Prescribed Minimum Cross-Sectional Area	Zero	Non-Zero but Small (10^{-8} to 10^{-12})

Layout optimization methods used by the numerical school (e.g. Kirsch 1989; Kirsch and Rozvany 1993) are usually based on the following two-stage procedure.

- First the topology is optimized for a given geometry (i.e. given coordinates of the joints), and then

- for this selected topology the geometry is optimized.

A drawback of this procedure is, of course, that for the new, optimized geometry the old topology may not be optimal any more. Until the introduction of COC/DCOC methods, however, the two-stage procedure was necessary because of the limited optimization capability of other methods, particularly for realistic problems with active stress constraints for a very large number of members in the structural universe.

The new optimality criteria methods (COC/DCOC), which were discussed in the first lecture, enable us to carry out a *simultaneous optimization of topology and geometry*, because the number of elements in the structural universe is either infinite (analytical methods) or very large (numerical COC/DCOC methods) and hence topological optimization achieves, in effect, also geometrical optimization.

The theory of optimal structural layouts has been covered extensively in the past, in particular in principal lectures at NATO meetings [Iowa 1980 (Rozvany 1981); Troia 1986 (Rozvany and Ong 1987); Edinburgh 1989 (Rozvany, Gollub and Zhou 1992); Berchtesgarden 1991 (Rozvany, Zhou and Gollub 1993); Sesimbra 1992 (Rozvany 1993)], and also in greater detail at a CISM course in Udine in 1990 (Rozvany 1992b) as well as in books (Rozvany 1976, 1989) and in book chapters (e.g. Rozvany 1984). Although the above exposure in the technical literature may appear excessive, the development in this field has also been both rapid and extensive, even during the last few months. In this lecture, an extended layout theory for the *exact layouts* of multiload, multipurpose structures is presented. Numerical, discretized layout solutions for a variety of design conditions (see Table 1) were discussed elsewere (e.g. Zhou and Rozvany 1991; Rozvany, Zhou, Birker and Sigmund 1993; Rozvany, Zhou and Sigmund 1994).

3 An Extended Layout Theory for Multiload, Multipurpose Structures: Exact Analytical Solutions

3.1 Fundamental Features

The general formulation given here can be applied to any combination of design constraints for multiple load conditions, but will be discussed here in the context of either multiple displace constraints or stress constraints.

A new unified formulation of the structural layout theory is presented in this section. It was mentioned in Table 1 that for layout problems we consider a structural universe with an infinite number of members, that is, with members running in all possible directions at all points of the available space. Moreover, any cross-sectional area can take on a zero value. The difference between this class of problems and cross-section optimization by DCOC is that in the latter case the number of elements is finite and a lower limit is imposed on cross-sectional areas.

The *proof* of the optimality conditions listed in this section can be based on *energy theorems* or *calculus of variations* but can also be derived from the *standard optimality conditions of DCOC*. In doing so, the following two limiting processes are employed:

- the lower limit on cross-sectional parameters tends to zero: $x_l^e \to 0$, and

- the number of members passing through a point of the available space tends to infinity.

It follows from the first of the above operations that for a zero cross-section we have an inequality in the optimality criterion (as in the case of $x^e = x_l^e$ in DCOC).

The second step above implies, due to the kinematic admissibility requirement of the adjoint strains in DCOC, that we can replace separate adjoint strains in bars running in all possible directions at a point with a single strain field (for plane trusses, for example, with a single plane strain field). Some mathematicians prefer avoiding the above transitions. For example, in a mathematically outstanding contribution, Strang and Kohn (1983) describe the "Michell truss"

140

problem as minimizing the integral $\int \int (|\sigma_1| + |\sigma_2|)\, dx dy$ for a plane stress field. Michell's original problem would reduce to the Strang-Kohn problem after

- making the second transition above; and

- showing that the optimal member directions must coincide with the principal stress (and strain) directions.

The above steps are actually part of the standard optimal layout theory (e.g. Prager and Rozvany 1977a and b).

For all layout problems considered by the authors, the fundamental layout optimality criterion can be stated in terms of a *"criterion function"* ϕ^e for a given element e in the following form (Rozvany 1992a)

$$\phi^e = 1 \ \ (\text{for } A^e \neq 0), \quad \phi^e \leq 1 \ \ (\text{for } A^e = 0), \tag{1}$$

where A^e is the cross-sectional area of an element e. The element e as well as the criterion function f^e (and A^e) are location and direction dependent, that is, $\phi^e = \phi^e(x, y, \theta)$ in the plane and $\phi^e = \phi^e(x, y, z, \theta, \eta)$ in 3D space where (x, y, z) are spatial coordinates and (θ, η) define the orientation of a line element.

3.2 Applications to a Simple Class of Structures

For illustration purposes, we consider here *grid-type* structures with one design variable x^e and one nodal force f^e per element. Moreover, we assume that the element cost (weight) can be expressed as $c^e L^e x^e$, where c^e is a given constant and L^e is the length of the element, and the element stiffness takes the form $r^e x^e / L^e$, where r^e is a given constant. For *trusses*, for example, x^e is the cross-sectional area $(x^e := A^e)$, c^e is the specific weight $(c^e := \rho^e)$ and r^e is Young's modulus $(r^e := E^e)$. For *grillages* with beams of given depth d^e but variable width x^e, we have $A^e := x^e d^e$, $c^e := d^e \rho^e$, $r^e := E^e (d^e)^3 / 12$.

The *generalized strains* ε^e used below mean axial strain (usually ε^e) for trusses and curvatures $(\varepsilon^e := \kappa^e)$ for grillages. Moreover, the *generalized forces* f^e refer to member forces (previously f^e) in the case of trusses and to bending moments $(f^e := M^e)$ in the case of grillages.

Stress constraints can be formulated as $(|f^e|/a^e - x^e) \leq 0$, in which for trusses $a^e := \sigma^e$ and for grillages of given depth $a^e := \sigma^e (d^e)^2 / 6$, where σ^e is the permissible stress. For *plastic design*, $a^e = \sigma^e (d^e)^2 / 4$ and σ^e is the yield stress.

For various classes of *design conditions* the criterion functions are given below.

3.3 Criterion Functions for Displacement Constraints

(a) *Structures with several load conditions $(k = 1, \ldots, K)$ and multiple displacement constraints $(j = 1, \ldots, J_k)$.* For this class of problems, the criterion function becomes (Rozvany 1992a)

$$\phi^e = (r^e / c^e) \sum_{k=1}^{K} (\varepsilon_k^e \overline{\varepsilon}_k^e), \tag{2}$$

with

$$\varepsilon_k^e = \frac{f_k^e}{r^e x^e}, \quad \overline{\varepsilon}_k^e = \frac{\sum_{j=1}^{J_k} \nu_{kj} \hat{f}_{kj}^e}{r^e x^e} = \frac{\overline{f}_k^e}{r^e x^e}, \tag{3}$$

where ε_k^e and $\overline{\varepsilon}_k^e$ are the real and adjoint strains for the element e and load condition k, f_k^e the corresponding generalized forces, $\hat{f}_{kj}^e$ is the virtual generalized force for load k and displacement j, $\overline{f}_k^e$ is the adjoint generalized force for the load and ν_{kj} the Lagrange multiplier for the load k and displacement constraint j. The value of ν_{kj} is nonzero only if the j-th displacement condition for the k-th load is active, that is, it is satisfied as an equality. For the optimal design variable x^e, we have from COC/DCOC (e.g. Rozvany 1992a; Rozvany and Zhou 1994):

$$x^e = \sqrt{\frac{\sum_k f_k^e \sum_j \nu_{kj} \hat{f}_{kj}^e}{r^e c^e}} = \sqrt{\frac{\sum_k f_k^e \overline{f}_k^e}{r^e c^e}}. \tag{4}$$

(b) *Structures with proportional displacement constraints.* These constraints restrict a weighted combination of displacements at the location and in the direction of point loads, with weighting factors proportional to the magnitude of the loads. For a single point load, this means constraints on a single displacement at and in the direction of that load. Proportional displacement constraints can be transformed into a *compliance constraint*, restricting the *total external work*.

Considering *structures with several load conditions* $(k = 1, \ldots, K)$ and a *proportional displacement constraint or compliance constraint* for each load condition, we have $f_k^e = \hat{f}_k^e$, $\overline{f}_k^e = \nu f_k^e$, $\overline{\varepsilon}_k^e = \nu_k \varepsilon_k^e$ and hence (2)–(4) reduce to

$$\phi^e = (r^e/c^e) \sum_{k=1}^{K} \nu_k (\varepsilon_k^e)^2 , \tag{5}$$

$$\varepsilon_k^e = \frac{f_k^e}{r^e x^e}, \quad \overline{\varepsilon}_k^e = \frac{\nu f_k^e}{r^e x^e}, \tag{6}$$

$$x^e = \sqrt{\frac{\sum_k \nu_k (f_k^e)^2}{r^e c^e}}. \tag{7}$$

(c) *Structures with one single proportional displacement constraint or compliance constraint (one load condition).* In this case, (5) and (7) reduce to

$$\phi^e = (r^e/c^e)\nu(\varepsilon^e)^2 , \tag{8}$$

$$x^e = \sqrt{\frac{\nu}{r^e c^e}} |f^e|. \tag{9}$$

3.4 Criterion Functions for Stress Constraints

(a) *Structures with several load conditions and stress constraints.* For this problem, we have

$$\phi^e = (a^e/c^e) \sum_{k=1}^{K} (\overline{\varepsilon}_k^e \, \mathrm{sgn} \, f_k^e) \,, \tag{10}$$

where the meaning of sgn f_k^e is sgn $f_k^e = 1$ for $f_k^e > 0$, sgn $f_k^e = -1$ for $f_k^e < 0$ and $(-1) \leq \mathrm{sgn} \, f_k^e \leq 1$ for $f_k^e = 0$. Since for $A^e = 0$ we have $f_k^e = 0$ (for all k) and (10) must be fulfilled for any values of the "sgn" functions, for $A^e = 0$ (1) and (10) reduce to

$$\phi^e = (a^e/c^e) \sum_{k=1}^{K} |\overline{\varepsilon}_k^e| \leq 1 \quad (\text{for } A^e = 0) \,, \quad \overline{\varepsilon}_k^e \neq 0 \text{ only if } |f_k^e|/a^e = x^e \,. \tag{11}$$

Important Note. In the derivation of DCOC, only statical admissibility is assumed (Zhou and Rozvany 1992) and kinematic admissibility becomes an optimality criterion *if a displacement constraint is active.* Since in the considered class of problems no displacement constraint is involved, the optimality conditions in (1) with (10) represent an optimal solution only *if the strains ε_k^e in nonvanishing members are kinematically admissible.* This condition is always satisfied for statically determinate solutions, which is the case for a single load condition (ignoring non-unique optima) and sometimes for several load conditions. However, in *optimal plastic design* no kinematical admissibility of the real structure is required and hence a solution satisfying (1) and (10) always represents (in convex problems) the optimum.

(b) *Structures with one load condition and stress constraints.* In this case the criterion function becomes

$$\phi^e = (a^e/c^e) \overline{\varepsilon}^e \, \mathrm{sgn} \, f^e \,, \tag{12}$$

and then (1) and (12) imply

$$\overline{\varepsilon}^e = (c^e/a^e) \, \mathrm{sgn} \, f^e \quad (\text{for } A \neq 0) \,, \quad |\overline{\varepsilon}^e| \leq c^e/a^e \quad (\text{for } A = 0) \,. \tag{13}$$

For trusses with $c^e = \mathrm{const.}$, $a^e := \sigma^e$ (13) reduces to the optimality conditions of Michell (1904). For grillages, (13) reduces to optimality conditions of Rozvany (1972; for a review see Prager and Rozvany 1977a).

3.5 General Procedure for Exact Layout Optimization

(a) Adopt continuum type strain fields ε_k^e and $\overline{\varepsilon}_k^e$ covering the entire available space and satisfying kinematic boundary and continuity conditions, such that
(b) the criterion function ϕ^e takes on a directionally maximum value of $\phi^e = 1$ at each point.

(c) Place members along lines with $\phi^e = 1$, calculate the member forces and dimension the members using optimality criteria for the cross-sectional parameters [e.g. for x^e in (4)].

(d) Check that the strains and adjoint strains in nonvanishing members (with $A^e \neq 0$) equal those given by the strain fields under (a).

Naturally, the above operations must be carried out simultaneously because all the conditions above must be fulfilled by the solution.

3.6. Self-Adjoint Problems

(a) One loading Condition. If we square both sides of (13) and adopt $(\phi^e)^2$ as a modified criterion function, we can see that structures of the type considered here (e.g. plane trusses, and grillages of given depth) have the *same optimal layout for one load condition and either a proportional displacement (or compliance constraint) or a stress constraint.*

Moreover, it follows from (14) that for one loading condition, *the optimal member directions must coincide with the principal directions of the adjoint strain field.*

This property was pointed out, in the context of trusses, already by Michell (1904) and was used by Hemp (1973) and Prager (1974) for deriving some *least-weight plane truss layouts for one load condition.* A more comprehensive treatment of the above class of problems was given by Rozvany and Gollub (1990), Lewiński, Zhou and Rozvany (1993, 1994) and Rozvany, Lewiński *et al.* (1993). For numerical examples, see the above references.

The *optimal layout of grillages for one load condition* has been discussed extensively in the literature (e.g. Rozvany 1972; Prager 1974; Prager and Rozvany 1977a; Rozvany 1976, 1981, 1984, 1989). The latest extension treated allowance for the cost of supports (Rozvany 1994).

(b) Several Loading Conditions. In the case of proportional displacement constraints or compliance constraints, optimal layouts have been determined for plane trusses with several load conditions (Rozvany, Zhou and Birker 1993; Rozvany, Birker and Lewiński 1994). It has been found that, even in the case of nonproportional displacement constraints, at any point of the optimal layout for plane trusses, *members may run in at most two directions,* which are *in general non-orthogonal.* A theory for *non-orthogonal, generalized Hencky-nets* for the layout of plane trusses with several load conditions has also been developed (Rozvany and Birker 1994). Numerical examples can be found in the above references.

3.7 Non-Self-Adjoint Problems

The above result have been extended to problems with nonproportional displacement constraint, which are in general non-self-adjoint and non-convex (e.g. Rozvany, Sigmund, Lewiński *et al.* 1993). It was found that

- due to nonuniqueness, several solutions satisfying the optimality critria may need to be investigated;

- the global optimal solution may be nonstationary if too few displacement constraints are prescribed; and

- the solution given by the optimality criteria may represent a local maximum (and not a local minimum).

4 Concluding Remarks

In this paper, an extended layout theory for deriving *exact, analytical solutions* for multipurpose, multiload structures was presented. These solutions are useful in checking the validity and convergence of the discretized numerical methods presented in this paper. The alternative method of optimizing the topology for compliance only, and then the geometry for a combination of design conditions, has been shown to lead to grossly nonoptimal solutions (e.g. Sankaranarayanan, Haftka *et al.* 1992).

6 Acknowledgements

With regards to both lectures published herein, the authors are indebted to the Deutsche Forschungsgemeinschaft and the Humboldt Foundation for financial support, to Peter Moche (drafting), to Sabine Liebermann (assembling, checking and processing the text) and to Susann Rozvany (editing).

6 References

Hemp, W.S. 1973: *Optimum Structures.* Clarendon, Oxford.

Kirsch, U. 1989: Optimal topologies of structures. *Appl. Mech. Rev.* **42**, 223–238.

Kirsch, U.; Rozvany, G.I.N. 1993: Design considerations in the optimization of structural topologies. In: Rozvany, G.I.N. (Ed.) *Optimization of Large Structural Systems* (Proc. NATO ASI held in Berchtesgaden, 1991), pp. 121–141, Kluwer, Dordrecht.

Lewiński, T.; Zhou, M.; Rozvany, G.I.N. 1993: Exact least-weight truss layouts for rectangular domains with various support conditions. *Struct. Optim.* **6**, 65–67.

Lewiński, T.; Zhou, M.; Rozvany, G.I.N. 1994: Extended exact solutions for least-weight truss layouts – Part I: Cantilever with a horizontal axis of symmetry – Part II: Unsymmetric cantilevers. *Int. J. Mech. Sci.* (accepted).

Michell, A.G.M. 1904: The limits of economy of material in frame-structures. *Phil. Mag.* **8**, 589–597.

Prager, W. 1974: *Introduction to Structural Optimization*. (Course held in Int. Centre for Mech. Sci. Udine. CISM **212**). Springer-Verlag, Vienna.

Prager, W.; Rozvany, G.I.N. 1977a: Optimal layout of grillages. *J. Struct. Mech.* 5, 1–18.

Prager, W.; Rozvany, G.I.N. 1977b: Optimization of structural geometry. In: Bednarek, A.R.; Cesari, L. (Eds.) *Dynamical Systems*, pp. 265–293. Academic Press, New York.

Rozvany, G.I.N. 1972: Grillages of maximum strenght and maximum stiffness. *Int. J. Mech. Sci.* **14**, 651–666.

Rozvany, G.I.N. 1976: *Optimal Design of Flexural Systems*. Pergamon Press, Oxford. Russian translation: Stroiizdat, Moscow, 1980.

Rozvany, G.I.N. 1981: Optimality criteria for grids, shells and arches. In: Haug E.J.; Cea, J. (Eds.) *Optimization of Distributed Parameter Structures* (Proc. NATO ASI held in Iowa City, 1980), pp. 112–151. Sijthoff and Noordhoff, Alphen aan der Rijn, The Netherlands.

Rozvany, G.I.N. 1984: Structural layout theory: the present state of knowledge. In: Atrek, E.; Gallagher, R.H.; Ragsdell, K.M.; Zienkiewicz, O.C. (Eds.) *New Directions in Optimum Structural Design*, pp. 167–195. Wiley & Sons, Chichester, England.

Rozvany, G.I.N. 1989: *Structural Design via Optimality Criteria*, Kluwer, Dordrecht.

Rozvany, G.I.N.; 1992a: Optimal layout theory: analytical solutions for elastic structures with several deflection constraints and load conditions. *Struct. Optim.* 4, 247–249.

Rozvany, G.I.N. 1992b: Optimal Layout Theory (Chapter 6, see also Chapter 7-10). In: Rozvany, G.I.N. (Ed.) *Shape and Layout Optimization of Structural Systems and Optimality Criteria Methods* (CISM Course held in Udine 1990), pp. 75–163, Springer-Verlag, Vienna.

Rozvany, G.I.N. 1993: Layout theory for grid-type structures. In: Bendsøe, M.P.; Mota Soares, C.A. *Topology Design of Structures* (Proc. NATO ARW. Sesimbra 1992), pp. 251–272. Kluwer, Dordrecht

Rozvany, G.I.N. 1994: Optimal layout theory – allowance for the cost of supports and optimization of support locations. *Mech. Struct. Mach.* (proofs returned).

Rozvany, G.I.N.; Birker, T. 1994: Optimal truss layouts for multiple load conditions via generalized Hencky nets. *Struct. Optim.* (submitted).

Rozvany, G.I.N.; Birker, T.; Lewiński, T. 1994: Some unexpected properties of exact least-weight plane truss layouts with displacement constraints for several alternate loads. *Struct. Optim.* **7**, (scheduled for No. 1)

Rozvany, G.I.N.; Gollub, W. 1990: Michell layouts for various combinations of line supports, Part I. *Int. J. Mech. Sci.* **32**, 12, 1021–1043.

Rozvany, G.I.N.; Gollub, W.; Zhou, M. 1992: Layout optimization in structural design. In: Topping, B.H.V. (Ed.) *Proc. NATO ASI, Optimization and Decision*

Support Systems in Civil Engeneering, held 25 June – 7 July 1989, Edinburgh. Kluwer, Dordrecht.

Rozvany, G.I.N.; Lewiński, T.; Gerdes, D.; Birker, T. 1993: Optimal topology of trusses or perforated deep beams with rotational restraints at both ends. *Struct. Optim.* **5**, 268–270.

Rozvany, G.I.N.; Olhoff, N.; Bendsøe, M.P.; Ong, T.G.; Sandler, R.; Szeto, W.T. 1987: Least-weight design of perforated elastic plates I, II. *Int. J. Solids Struct.* **23**, 521–536, 537–550.

Rozvany, G.I.N.; Olhoff, N.; Cheng, K.-T.; Taylor, J.E. 1982: On the solid plate paradox in structural optimization. *J. Struct. Mech.* **10**, 1–32.

Rozvany, G.I.N.; Ong, T.G. 1987: Minimum-weight plate design via Prager's layout theory (Prager memorial lecture). In: Mota Soares (Ed.) *Computer Aided Optimal Design: Structural and Mechanical Systems* (Proc. NATO ASI held in Troia, Portugal, 1986), pp. 165–179. Springer-Verlag, Berlin.

Rozvany, G.I.N.; Sigmund, O.; Lewiński, T.; Gerdes, D.; Birker, T. 1993: Exact optimal structural layouts for non-self-adjoint problems. *Struct. Optim.* **5**, 204–206.

Rozvany, G.I.N.; Zhou, M. 1991: Applications of the COC method in layout optimization. In: Eschenauer, H.; Matteck, C.; Olhoff, N. (Eds.) *Proc. Int. Conf. "Engineering Optimization in Design Processes"* (Karlsruhe 1990), pp. 59–70, Springer, Berlin.

Rozvany, G.I.N.; Zhou, M. 1994: Optimality Criteria Methods for Large Structural Systems. Chaper 2 in: Adeli, H. (Ed.) *Advances in Design Optimization.* Chapman and Hall, London.

Rozvany, G.I.N.; Zhou, M., Birker, T. 1992: Generalized shape optimization without homogenization. *Struct. Optim.* **4**, 250–252.

Rozvany, G.I.N.; Zhou, M., Birker, T. 1993: Why multi-load design based on orthogonal microstructures are in general non-optimal. *Struct. Optim.* **6**, 200–204.

Rozvany, G.I.N.; Zhou, M., Birker, T.; Sigmund, O. 1993: Topology optimization using iterative continuum-type optimality criteria (COC) methods for discretized systems. In: Bendsøe, M.P.; Mota Soares, C.A. *Topology Design of Structures. Proc. NATO ARW.* (held in Sesimbra 1992)

Rozvany, G.I.N.; Zhou, M.; Gollub, W. 1993: Layout optimization by COC methods: analytical solutions. In: Rozvany, G.I.N. (Ed.) *Optimization of Large Structural Systems* (Proc. NATO ASI held in Berchtesgaden, 1991), pp. 77–102, Kluwer, Dordrecht.

Rozvany, G.I.N.; Zhou, M.; Sigmund, O. 1994: Optimization of topology. Chapter 10 in Adeli, H. (Ed.) *Advances in Design Optimization*, Chapman and Hall, London.

Sankaranarayanan, S.; Haftka, R.T.; Kapania, R.K. 1992: Truss topology optimization with simultaneous analysis and design. *Proc. 33rd AIAA/ASME/*

ASCE/AHS/ASC Struct. Dyn. Mat. Conf. (held in Dallas), pp. 2576–2585. AIAA, Washington DC.

Strang, G.; Kohn, R.V. 1983: Hencky-Prandtl nets and constrained Michell trusses. *Comp. Meth. Appl. Mech. Eng.* **36**, 207–222

Zhou, M.; Rozvany, G.I.N. 1991: The COC algorithm, Part II: topological, geometrical and generalized shape optimization. *Comp. Meth. Appl. Eng.* **89**, 309–336.

Zhou, M.; Rozvany, G.I.N. 1992: DCOC: an optimality critria method for large systems, Part I: theory. *Struct. Optim.* **5**, 12–25.

Homology Design of Flexible Structure by the Finite-Element Method

N.Yoshikawa and S.Nakagiri

Institute of Industrial Science, University of Tokyo,
7-22-1, Roppongi, Minato-ku, Tokyo 106, Japan

Abstract. A new method of homology design based on finite element sensitivity analysis is proposed in static problems. The stiffness equation is separated into two equations by introducing homologous constraint. The governing equation of the design variables is derived from the condition for the two equations to hold. The Moore-Penrose generalized inverse is employed to determine the design variables, as the governing equation is expressed in a rectangular matrix form. The validity of the proposed method is demonstrated by the numerical example concerning with out-of-plane bending of a planar frame structure, for which the homologous constraint is set to keep a member straight.

Keywords. Homology design, Moore-Penrose generalized inverse,
Optimum design, Finite element method,
Sensitivity analysis, Flexible structure

1. Introduction

One of the basic purposes of conventional structural optimization is to enhance the integrity of structure against external loads[1]. Sophisticated design methodology is required in the design of large flexible structures such as the space structure, where simultaneous optimum design of the structures and their control system to fulfill stringent demands is to be taken into consideration[2]. Homology design is expected to be a candidate of the optimum design for such structures. The methodology to utilize homologous deformation is called homology design in this paper. When a prescribed geometrical property at the whole or in particular part of a structure is kept before, during and after deformation, the deformation is called homologous[3]. The cost to control structural deformation mode would be mitigated by realizing some homologous deformation specified adequately corresponding to the demand to the structure. The performance of large radio telescope was greatly improved by applying homology design to shape control of tiltable parabolic antenna[3][4]. Hangai proposed a method of homology design by utilizing finite element method, which is based on the existence condition of the solution for modified stiffness equation and requires computation of derivative of generalized inverse, resulting in long CPU time[5].

As a kind of finite element synthesis, a formulation is presented herein for the homology design in static problems. The formulation is different from the Hangai's method, and is based on the separation of the stiffness equation and the finite element sensitivity analysis. The validity and versatility of the proposed formulation is demonstrated through the numerical examples in problems of out-of-plane bending of a planar frame structure. The homologous deformation to keep a central member of the structure straight is realized by changing the diameter of cross section or nodal position of the structure in the examples.

2. Governing Equation for Homology Design

The stiffness equation of a linear structure by the baseline design after the incorporation of the geometrical boundary conditions is given in the form of eq.(1) by finite element discretization in static problems,

$$[K]\{u\} = \{f\} \tag{1}$$

where $[K]$ is the stiffness matrix, $\{u\}$ the unknown nodal displacement vector and $\{f\}$ the external load vector. The displacement vector of N ingredients is partitioned into two parts due to the constraint of homologous deformation. The first part of it is denoted by $\{u_i\}$ with N-J ingredients. The second part $\{u_h\}$ consists of J ingredients subjected to the homologous constraint.

$$\{u\} = \begin{Bmatrix} u_i \\ u_h \end{Bmatrix} \tag{2}$$

The constraint of homologous deformation is dealt with in the form of eq.(3) through constraint $[C]$ matrix in this study.

$$\{u_h\} = [C]\{u_i\} \tag{3}$$

The stiffness equation of eq.(1) is partitioned as eq.(4) corresponding to the partitioning of the displacement vector of eq.(2).

$$\begin{bmatrix} K_{ii} & K_{ih} \\ K_{hi} & K_{hh} \end{bmatrix} \begin{Bmatrix} u_i \\ u_h \end{Bmatrix} = \begin{Bmatrix} f_i \\ f_h \end{Bmatrix} \tag{4}$$

150

Equation (4) is separated into two eqs.(5) and (6) when the expression of eqs.(7) and (8) is employed.

$$[K_s]\{u_i\} = \{f_i\} \tag{5}$$

$$[K_r]\{u_i\} = \{f_h\} \tag{6}$$

$$[K_s] = [K_{ii}] + [K_{ih}][C] \tag{7}$$

$$[K_r] = [K_{hi}] + [K_{hh}][C] \tag{8}$$

$[K_s]$ is a $(N\text{-}J)\text{x}(N\text{-}J)$ square and asymmetric matrix, and $[K_r]$ is a $J\text{x}(N\text{-}J)$ rectangular one. Equation (5) can be solved in the usual manner with respect to the square and nonsingular matrix $[K_s]$. If eq.(6) holds after substituting the solution $\{u_i\}$ of eq.(5), the objective homologous deformation can be said realized *per se*. In most case, eq.(6) is not satisfied. Hence, the baseline design is to be changed so as to eqs.(5) and (6) are satisfied by the same $\{u_i\}$. In doing so, M structural parameters P_m are chosen judiciously, and design variables α_m are assigned to them in the form of eq.(9) for the homologous deformation realization.

$$P_m = \overline{P}_m(1 + \alpha_m) \tag{9}$$

The upper bar means the quantities concerning with baseline design hereafter.

First-order approximation method is employed to decide α_m. The change of $\{u_i\}$ is expressed in the first-order approximation form of eq.(10) in the vicinity of the baseline design.

$$\{u_i\} = \{\overline{u}_i\} + \sum_{m=1}^{M}\{u_{im}^I\}\alpha_m \tag{10}$$

The superfix I and suffix m mean the first-order sensitivity with respect to α_m hereafter. $\{u_{im}^I\}$ is determined so that eq.(5) holds regardless of the design change. The change of the matrix $[K_r]$ is similarly expressed in the form of eq.(11).

$$[K_r] = [\overline{K}_r] + \sum_{m=1}^{M}[K_{rm}^I]\alpha_m \tag{11}$$

The governing equation of α_m is obtained as eq.(12) by means of substituting eqs.(10) and (11) into eq.(6), which must hold after the design change, and truncating second-order term of α_m.

$$\sum_{m=1}^{M} \left(\left[K'_{rm} \right] \{ \bar{u}_i \} + \left[\bar{K}_r \right] \{ u'_{im} \} \right) \alpha_m = \{ f_h \} - \left[\bar{K}_r \right] \{ \bar{u}_i \} \tag{12}$$

Equation (12) is summarized in the following matrix form,

$$[A]\{\alpha\} = \{b\} \tag{13}$$

where $[A]$ is a JxM rectangular matrix and $\{\alpha\}$ a design variable vector consisting of M unknowns. As $\{\bar{u}_i\}$ and $\{u'_{im}\}$ are proportional to $\{f_i\}$, the design variables are governed by the mode of loading, irrespective of the magnitude of it.

The solution of eq.(13) is obtained by using the Moore-Penrose generalized inverse $[A]^-$. When the existence condition of the solution in the form of eq.(14) holds, the general solution of eq.(13) is given as eq.(15),

$$\left([I] - [A][A]^- \right) \{b\} = \{0\} \tag{14}$$

$$\{\alpha\} = [A]^- \{b\} + \left([I] - [A]^- [A] \right) \{h\} \tag{15}$$

where $[I]$ is an identity matrix and $\{h\}$ an arbitrary vector consisting of M ingredients[6]. The first term and the second one of right hand side of eq.(15) are called particular solution and complementary solution, respectively. Aiming at minimum design change, we employ only the particular solution, which is so determined as to minimize the norm of the design variable vector, owing to the property of the Moore-Penrose generalized inverse.

The design variables thus obtained are affected by the deficiency of the first-order approximation. The deficiency can be overcome by iterative renewal of the baseline design. For the judgment of the realization of homologous deformation, the norm of vectorial difference between the achieved $\{u_h\}$ and the objective $\{u_h\}=[C]\{u_i\}$, as is given by eq.(16), is evaluated in this study.

$$\{\varepsilon\} = \{u_h\} - [C]\{u_i\} \tag{16}$$

The renewal of the design is carried out until the norm of vectorial error $\{\varepsilon\}$ becomes equal to nil satisfactorily.

3. Numerical Examples

Validity of the proposed formulation is examined in problems of out-of-plane bending of the planar frame structure illustrated in Fig.1. It is simply supported at

152

four corners marked by solid triangle and subjected to a concentrated force $p = 500$ N at the center node. The cross section of all members is kept circular whose diameter is equal to $d = 50.0$ mm for the initial baseline design. Each member is represented by a finite beam element with torsion. Young's modulus and modulus of rigidity of the material are taken as $E = 70.0$ GPa and $G = 26.9$ GPa, respectively. The deformation at the initial baseline design is shown in Fig.2.

The homologous deformation aimed at in this example is to keep the center member, which is indicated by s-s and encircled by the broken line in Fig.1, straight always. This example is associated with a scenario that the center member mounted with laser optical system is to be kept straight notwithstanding the disturbance load of system weight. The control cost to maintain the straightness would be lowered substantially by the realization of the homologous deformation. Design change is carried out in two ways, that is, by changing the diameter of all elements and by changing the nodal position of the structure. The error index for homologous deformation β is defined as the norm of the vectorial error normalized by the vertical deflection at center node denoted by w_s in the form of eq.(17) in this study.

$$\beta = \|\varepsilon\| / |w_s| \tag{17}$$

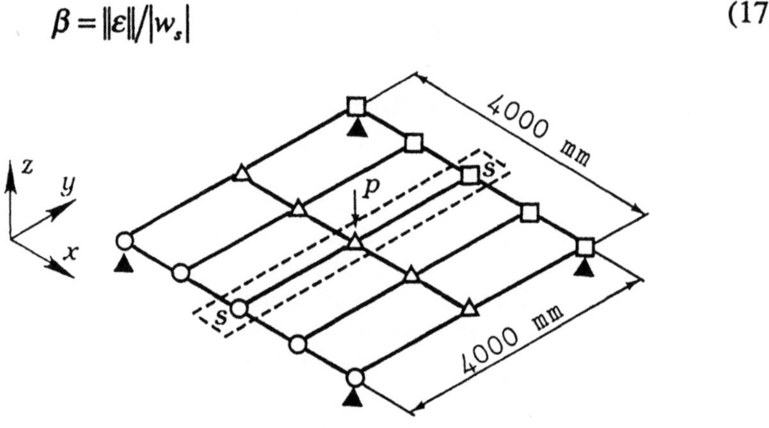

Fig.1 Planar frame structure

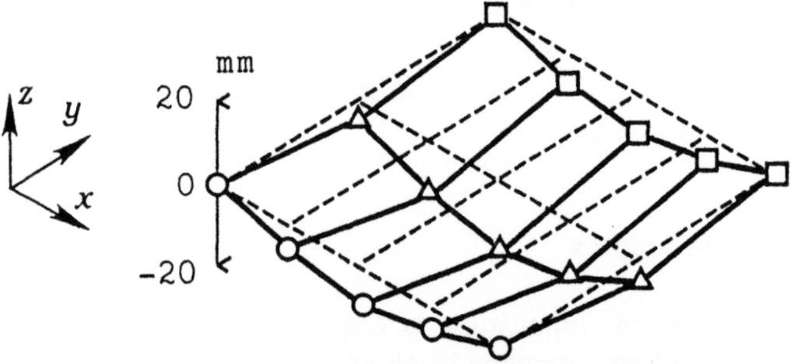

Fig.2 Deformation at baseline design

3.1 Homologous Deformation by Changing the Diameters

The effect of the diametral change is discussed first. Figure 3 shows the deformed frame after the first renewal, which is almost the same with that in Fig.2. Figure 4 depicts the deformed frame after the fifth renewal, in which the center member is kept straight as is aimed at. The amount of the change of the diameter to realize the objective homologous deformation is shown as a ratio to the initial value in Figs.5 and 6 for the members arrayed along the x direction and y direction, respectively. A simple way to realize the objective homologous deformation is to set the flexural rigidity of the center member sufficiently large, which leads to inevitable increase of total weight. The results of Figs.5 and 6 indicate that the objective homologous deformation is realized by increasing the diameter by 50% at most and total weight by 12.4% of initial baseline design. This subtle change in total weight owes to the property of the particular solution of the design variable vector whose norm is minimum. Figure 7 shows the monotonic convergence of the error index for homologous deformation to zero with iterative renewal.

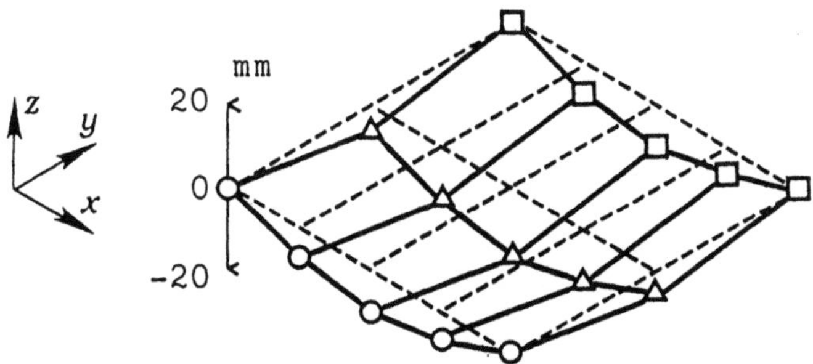

Fig.3 Deformation after first renewal in diameter

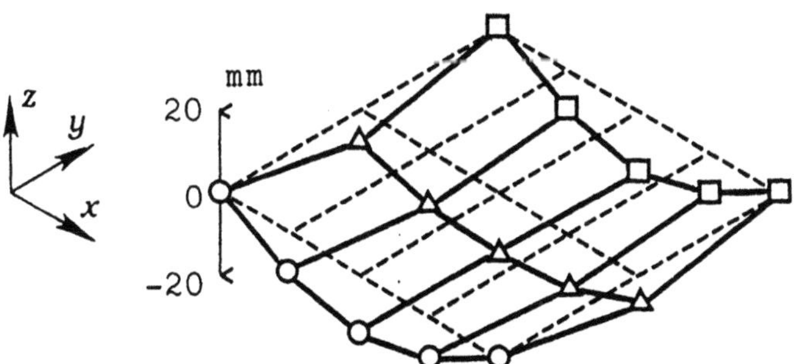

Fig.4 Homologous deformation realized by changing the diameter

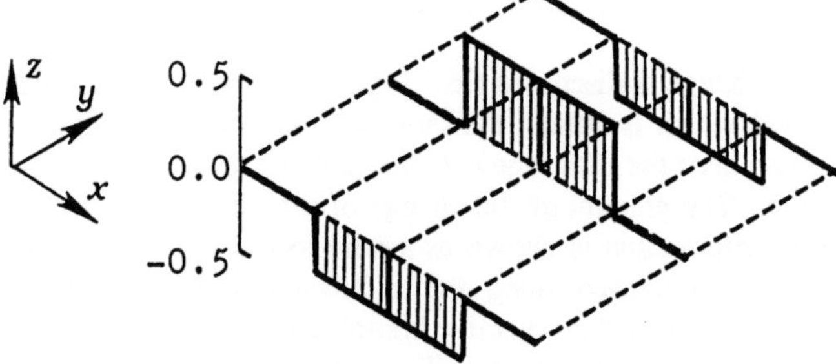

Fig.5 Amount of design change in diameter (*x* direction)

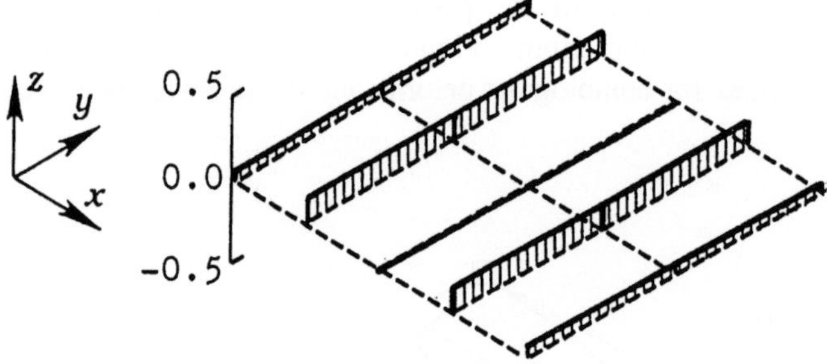

Fig.6 Amount of design change in diameter (*y* direction)

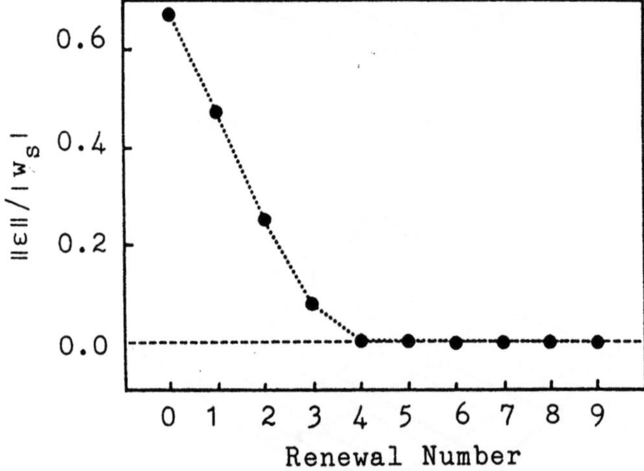

Fig.7 Convergence of error index by changing diameter

3.2 Homologous Deformation by Changing the Nodal Position

The objective homologous deformation is aimed at by changing the shape of the structure, that is, the nodal position. The design change in this case is carried out by assigning design variables independently to the length of members arrayed along y direction. Figure 8 shows the deformed frame viewed in the y direction, and indicates that sagging is left after the first renewal. Because of the symmetry of the deformation, the nodes marked by blank square in Fig.1 are not plotted. Nine renewals is required to realize the objective homologous deformation shown in Fig.9 in this case. The shape of the structure after the realization is shown in Fig.10 as viewed in the z direction. It is inferred qualitatively that a simple way to realize the objective homologous deformation is to make the center member sufficiently short. The result of Fig.10 indicates that the center member is lengthened conversely for the realization. Homology design in this case makes structure so flexible that the vertical deflection of center node is increased remarkably. If demanded, some constraint condition for the overall stiffness to maintain the structural integrity should be added to the governing equation of the design variables. Figure 11 depicts that the convergence of the error index for homologous deformation is also monotonic in this case.

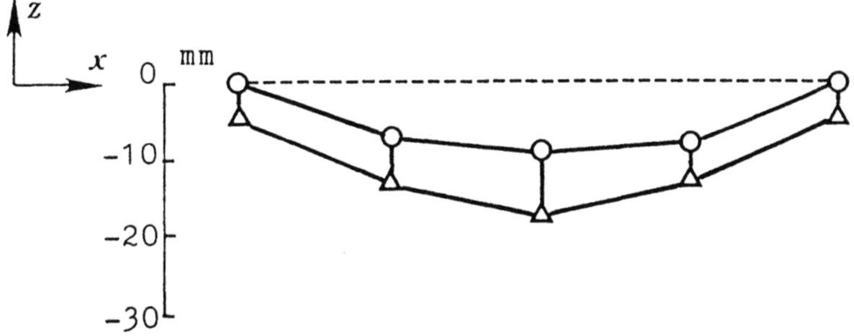

Fig.8 Deformation after first renewal in nodal position

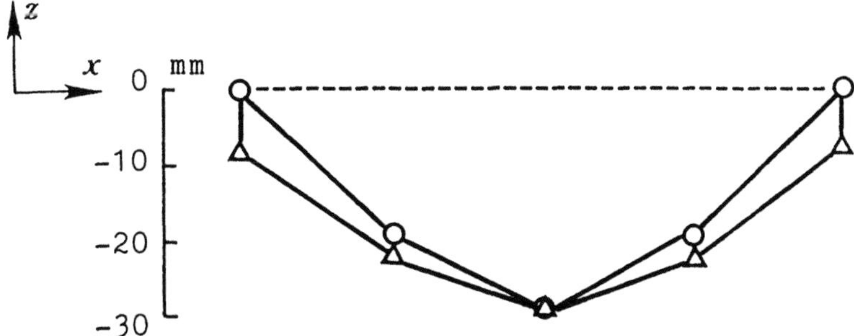

Fig.9 Homologous deformation realized by changing nodal position

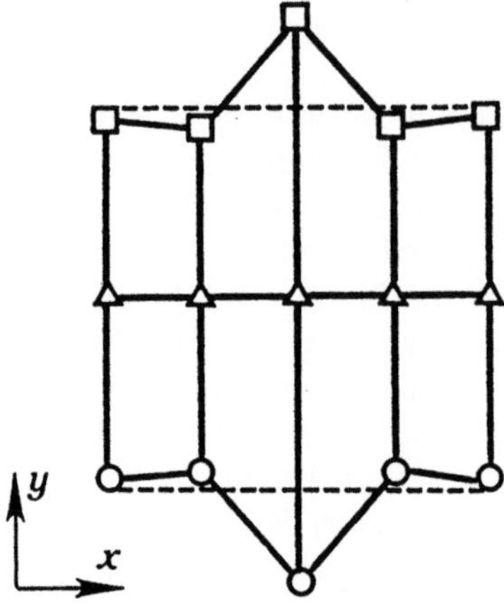

Fig.10 Top view of frame for homologous deformation

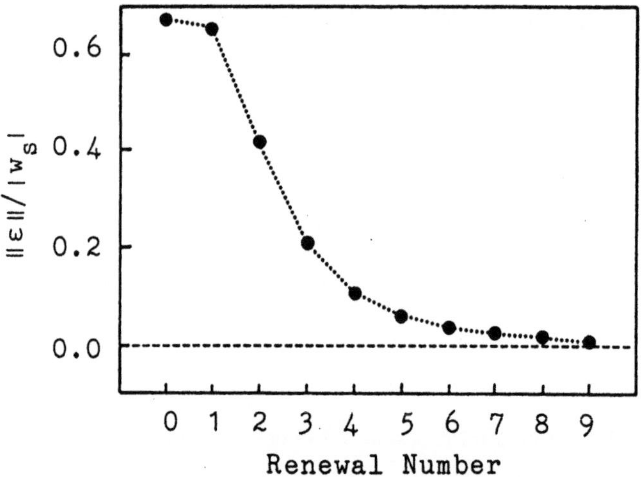

Fig.11 Convergence of error index by changing nodal position

4. Concluding Remarks

Homology design in static problems is investigated. The new methodology by means of finite element synthesis is proposed. The validity and versatility of the proposed method is examined in problems of out-of-plane bending of planar frame structure, in which the homologous deformation to keep the center member straight is realized by two ways of changing the diameter of all members and nodal position of the structure. The realization is carried out by changing structural parameters subtly, owing to the property of the design variables with the minimum norm. A numerical example suggests that homology design sometimes results in low stiffness structure which needs some constraint in order to maintain the structural integrity.

References

[1]Gallagher, R.H. and Zienkiewicz, O.C., (Eds.) : Optimum Structural Design-Theory and Applications. John Wiley & Sons, 1973.

[2]Grandhi, R.V., Haq, I. and Khot, N.S., : Enhanced Robustness in Integrated Structural/Control Systems Design. AIAA Journal, Vol.29, No.7, 1991, pp.1168-1173.

[3]Hoerner, S., : Homologous Deformations of Tiltable Telescopes. Journal of the Structural Division, Proceedings of ASCE, Vol.93, ST5, 1967, pp.461-485.

[4]Morimoto, M., Kaifu, N., Takizawa, Y., Aoki, K. and Sakakibara, O. : Homology Design of Large Antenna (in Japanese). Mitsubishi Electric Corp., Technical Report, Vol.56, No.7, 1982, pp.495-502.

[5]Hangai, Y. : Shape Analysis of Structures. Theoretical and Applied Mechanics, Vol.39, Proc. 39th Japan National Congress for Applied Mechanics, Ed. JNCTAM, Science council of Japan, Univ. of Tokyo Press, 1990, p.11-28.

[6]Rao, C.R. and Mitra, S.K., : Generalized Inverse of Matrices and Its Applications, John Wiley & Sons, 1971.

Optimization of Structure and Development of Production Systems

S. Shevchenko

Physical Engineering Dept., Kharkov Polytechnic Institute
Ul. (*street*) Frunze 21, Kharkov-2, 310002, UKRAINE

Abstract Problems encountered when developing some distributed production systems (or service systems), so as to make them capable of performing definite types of tasks are considered in the paper. These systems can be described by mathematical models used in the solution of problems of development and allocation of manufacturing tasks. Use is made of models describing events occurring at discrete time instants. The problem is reduced to an optimization of task distribution structure and to the choice of variants for the development of a system within the framework of allocated resources.

Keywords. Distributed systems, system development, manufacturing systems, discrete time, simulation, assignment problem, suboptimal solution, modified Lagrange function.

1. Problem formulation and mathematical models

There are some tasks, which are characterized by a set of required parameters at fixed time instants. All possible variants of the development with indication of cost and increase in parameter values at the considered time intervals have been prescribed for each manufacturing unit of the system (service center). The fulfilment of prescribed tasks requires some extra expenses resulting from the necessary interconnection between the task source and the service center. These expenses may be those of transcription costs, communication costs and etc.

One of the natural criteria in the solution of this problem is that the total expenses for the development of a service system and for the fulfilment of all tasks should be a minimum.

Another possible criterion may be that of considering the quality of functioning of the system as described by the downtime of the task sources waiting for the begining of servicing and also there may also be losses caused by equipment failure incurred at the servicing center and so on. The value of this criterion can be determined through a simulation modeling on the basis of the solution derived from the previous criterion.

Thus, we obtain a highly complex problem of optimization and simulation. Here, in the first stage, the structure of the servicing system and the task distribution are determined from the optimization problem. In the second stage the quality of the obtained solution is determined using the simulation model.

The equilibrium state of the criteria can be defined in the course of iterative interaction of the problems pointed out to by making use of the method of concessions and by changing the possible level of centralization in the system through a restriction of ultimate technical abilities of the service centers.

Let us consider the mathematical formulation of the optimization problem. Suppose, a set of adjustable time instants $T = \{t_1, t_2, ...\}$ has been prescribed. There are known the sets of tasks $I_t = \{i_1, i_2, i_3, ...\}$, $t \in T$, with the required characteristics of $\{w_{ik}\}$, $i \in I_t$, $k \in K$, where K is a multitude of types of such characteristics. Here J is a set of points selected for the location of service centers. For each point $j \in J$ the multitude of possible variants of the center development $\{P_{tj}\}$ has been prescribed in the form of the sets, reflecting the technical characteristics $\{B_{pjk}\}$ and the expenses required for the realization $\{A_{pj}\}$. The matrix of transformation expenses $\{c_{ijt}\}$ has been constructed on the multitudes $It \times J$.

The looked for variants of the system development and of the distribution of tasks between the service centers will be prescribed using the following logical variables

$$x_{ijt} = \begin{cases} 1 & \text{if i-th task source is handled in the} \\ & \text{j-th center at a time t,} \\ 0 & \text{otherwise} \end{cases}$$

$$y_{pjt} = \begin{cases} 1 & \text{if p-th variant of the development is used in} \\ & \text{the j-th center at a time t,} \\ 0 & \text{otherwise} \end{cases}$$

Then the mathematical model of this problem can be given in the following form:

$$S = \min \sum_t V_t [\sum_j [\sum_i c_{ijt} x_{ijt} + \sum_p [A_{pj} + g_{pj}(\Omega_{jt})] y_{pjt}]] \tag{1}$$

is to be found under the following constraints:

$$\sum_j x_{ijt} = 1, \quad i \in I_t, \ t \in T \tag{2}$$

$$\sum_i \omega_{ik} x_{ijt} \leq \sum_{\tau=1}^{t} \sum_p B_{pjk} y_{pj\tau}, \quad k \in K, \ j \in J, \ t \in T \tag{3}$$

$$\sum_i \sum_j c_{ijt} x_{ijt} \le c_t^0, \quad t \in T \tag{4}$$

$$\sum_j \sum_p A_{pj} y_{pjt} \le A_t^o, \quad t \in T \tag{5}$$

$$\sum_p y_{pjt} \le 1, \quad j \in J, \ t \in T \tag{6}$$

$$x_{ijt}, \ y_{pjt} \in \{0,1\}, \quad i \in I_t \ p \in P_{tj}, \ j \in J, \ t \in T. \tag{7}$$

Here Ω_{jt} is a vector, reflecting the total throughput of the j-th center according to the types of productions output:

$$\Omega_{jt} = \{\Omega_{kjt}\} = \{\sum_i \omega_{i1} x_{ijt}, \ \sum_i \omega_{i2} x_{ijt}, ...\},$$

$g_{pj}()$ is a function representing the current expenses in the j-th center according to the p-th variant of the fulfilment of the prescribed tasks, V_t - are the cost discounting coefficients.

The mathematical model (1)-(7) represents a problem of the integer non-linear programming with Boolean variables. The problems on development and location of production are considered to be known extensions of similar models.

Using the approach suggested in [1],[2] we shall partition the initial problem into a number of subproblems of the following type:

$$W_j(\xi_j) = \min \sum_t V_t [\sum_i c_{ijt} x_{ijt} + \sum_p [A_{pj} + g_{pj}(\Omega_{jt})] y_{pjt}] \tag{8}$$

is to be found under the constraints:

$$x_{ijt} = \xi_{ijt}, \quad i \in I_t, \ t \in T, \tag{9}$$

$$\sum_i \omega_{ik} x_{ijt} \le \sum_{\tau=1}^{t} \sum_p B_{pjk} y_{pj\tau}, \quad k \in K, \ t \in T, \tag{10}$$

$$\sum_p y_{pjt} \le 1, \quad t \in T, \tag{11}$$

$$x_{ijt}, \ y_{pjt} \in \{0,1\}, \quad i \in I_t, \ p \in P_{tj}, \ t \in T. \tag{12}$$

This problem describes the j-th service center. Its solution allows to determine the optimal variant of the center development for the prescribed distribution vector $\xi_j = \{\xi_{ijt}\}$. It also allows to find the subgradient components $\partial W_j(\xi_j) = \{-u_{ijt}\}$ of the function $S = \sum_j W_j(\xi_j)$ as the Lagrange factor values of constraints (9) in the stationary point of the Lagrange function of the subproblem taking into account the obtained solution. Then, the following coordinating problem gives an opportunity to change the existing distribution in the direction of the antisubgradient of the objective function of the initial problem

$$\max \sum_i \sum_j \sum_t u_{ijt}\xi_{ijt} \tag{13}$$

is to be found under the following constraints:

$$\sum_j \xi_{ijt} = 1, \quad i \in I_t, \ t \in T, \tag{14}$$

$$\sum_i \omega_{ik}\xi_{ijt} \le \sum_{\tau=1}^{t} \sum_p B_{pjk}\eta_{pj\tau}, \quad k \in K, \ j \in J, \ t \in T, \tag{15}$$

$$\sum_i \sum_j c_{ijt}\xi_{ijt} \le C_t^o, \quad t \in T, \tag{16}$$

$$\sum_j \sum_p A_{pj}\eta_{pjt} \le A_t^o, \quad t \in T, \tag{17}$$

$$\sum_p \eta_{pjt} \le 1, \quad j \in J, \ t \in T, \tag{18}$$

$$\xi_{ijt}, \ \eta_{pjt} \in \{0,1\}, \quad i \in I_t, \ p \in P_{tj}, \ j \in J, \ t \in T, \tag{19}$$

There, the variables ξ_{ijt} and η_{pjt} are the analog of the corresponding variables x_{ijt} and y_{pjt}. The coordinating problem represents a linear distribution problem with the Boolean variables or an assignment problem.

2. The Procedure of the solution of an assignment problem

Let us consider the problem of coordination (13)-(19). This problem is an NP-complete one. Therefore, the development of effective algorithms for fast derivation of suboptimal solutions is of special interest. We suggest such an approach to the solution of the problem considered as follows from the transformation of the initial problem using a modified Lagrange function. Let us construct the Lagrange function for the truncated problem (13)-(19).

$$\begin{aligned}
L \ = \ & \sum_i \sum_j \sum_t u_{ijt}\xi_{ijt} + \sum_j \sum_k \sum_t [\sum_{\tau=1}^{t} \sum_p B_{pjk}\eta_{pj\tau} + \\
& - \ \sum_i \omega_{ik}\xi_{ijt}]\lambda_{kjt} + \sum_t [[A_t^0 - \sum_j \sum_p A_{pj}\eta_{pjt}]\psi_t + [C_t^0 + \\
& + \ \sum_j \sum_i c_{ijt}\xi_{ijt}]\gamma_t] = \sum_t \sum_j \sum_i [u_{ijt} - c_{ijt}\gamma_t - \sum_k \omega_{ik}\lambda_{kjt}]\xi_{ijt} + \\
& + \ \sum_t \sum_j \sum_p [\sum_k B_{pjk}\sum_{\tau=t}^{T} \lambda_{kj\tau} - A_{pj}\psi_t]\eta_{pjt} + \sum_t [A_t\psi_t + C_t^0\gamma_t].
\end{aligned}$$

The constraints (14), (18), (19) prescribe the region where the function is to be determined.

The analysis of the obtained expression shows that if the optimal values of factors $\lambda = \{\{\lambda_{kjt}\}, \{\psi_t\}, \{\gamma_t\}\}$ then, owing to the action of constraints (14), (19) the unit values of variables ξ_{ijt}, appear in the optimal solution with the maximum coefficients under the unknown quantities for each $i \in I_t$, $t \in T$.

A similar conclusion can be drawn for the variables η_{pjt}, if the additional requirement of the positivity of these coefficients is met according to the condition (18).

However, the use of the Lagrange functions for discrete programming does not guarantee an optimal solution of the problem on account of duality discontinuity. Let us also note, that the main purpose of the coordinating problem is the determination of the task distribution among the service centers prescribed by the $\xi = \{\xi_{ijt}\}$. The variants of the development are being chosen in the subproblems along with the general estimation of this choice in the form of subgradient components of the objective function.

Moreover, the calculation experience shows that the constraints imposed on the total cost of equipment as well as those imposed on the total transportation expenditures are better to be taken into account according to the algorithm of intensities and also by introducing the corresponding weight factors for the arrays of $\{c_{ijt}\}$, $\{A_{pj}\}$ for the purpose of postoptimal analysis. Therefore, let us assume that $\psi_t = \gamma_t = 0$, $t \in T$. Then it is suggested to use the mentioned principles of variables assignment in the course of successive iterations using the following modification of the Lagrange function:

$$L_m = \sum_t \sum_j \sum_i [u_{ijt} - \lambda_{ijt}]\xi_{ijt} + \sum_t \sum_j \sum_p B_{pjk}\theta_{jt}\eta_{pjt}$$

here $\theta_{jt} = \sum_{\tau=t}^{T} \max_i \lambda_{ij\tau}$.

We shall start the search for the required distribution from the assignment formed under $\lambda_{ijt} = 0$, this corresponds to the constraint relaxation (15)-(17). Each subsequent assignment should differ from the preceding one by the value of one component, minimum change of the objective function and closer location to the admissible region. For this purpose we shall vary the introduced coefficients $\{\lambda_{ijt}\}$.

This will provide for the change of the assignment components with the same name, this assignment was obtained according to the regulations foreseen for the case when the corresponding constraints are not met.

Let us show the procedure of the choice of values for the main variables of the problem. Let us find

$$j^* = arg \max_j \{u_{ijt} - \lambda_{ijt}\}, \quad i \in I_t, \quad t \in T. \tag{20}$$

Then

$$\xi_{ijt} = \begin{cases} 1 & \text{if } j = j^* \\ 0 & \text{otherwise.} \end{cases} \tag{21}$$

Let's calculate θ_{jt}, $j \in J$, $t \in T$ and determine:

$$p^* = arg\max_{p} \sum_{k} B_{pjk}\theta_{jt}.$$

If $\sum_k B_{pjk}\theta_{jt} = 0$ for $p = p^*$ then $\forall p$: $\eta_{pjt} = 0$.
Otherwise

$$\eta_{pjt} = \begin{cases} 1 & \text{if } p = p^* \\ 0 & \text{otherwise.} \end{cases}$$

For the realization of the proposed algorithm it is necessary to form a search vector $\varphi^r = \{\varphi^r_{ijt}\}$ for each r-th iteration. This search vector prescribes the assignment coordinates which may be subject to change. The vector components φ can be prescribed in the following way:

$$\varphi_{ijt} = \begin{cases} 1 & \text{if } 3k \in K; \ \frac{\partial L}{\partial \lambda_{kjt}} < 0 \text{ and } \xi_{ijt} = 1, \\ 0 & \text{otherwise.} \end{cases}$$

We shall obtain new values of components $\lambda^{r+1} = \{\lambda^{r+1}_{ijt}\}$ for the next iteration. For this purpose we shall define the scalar value of $\rho^* = \min_{p>0}\rho$ under the condition that $\lambda^{r+1} = \lambda^r + \rho^*\varphi^r$ provides for the admissible change of the assignment of one order in terms of the expressions (20)-(21). If the next change of the distribution is inadmissible then λ^r is recalculated in terms of given relationship and the vector φ^r is redetermined and afterwards the process is continued.

The execution of this procedure allows to find the suboptimal solution of the problem (13)-(19) over a finite number of steps. This can be done owing to the minimum changes of the functional being optimized during the approximation to the admissible region. Let us note, that the components λ_{ijt} obtained using this approach carry the information about the conditional losses to the objective function in the course of assignment change relative to the initial value. If the iteration process will be continued after the derivation of admissible solution until only one assignment component remains unchanged, then we shall come to the inverse problem of the following type: to maximize the sum of indemnifield conventional losses by the restoration of the assignment components taking into account the constraints imposed on the productive capacities of the service centers.

Formally the inverse problem can be described by the following mathematical model.
Find

$$\mu(\xi^*) = \max\sum_{t}\sum_{j \in J'}\sum_{i}\lambda_{ijt}\xi_{ijt} \tag{22}$$

under the constraints

$$\sum_i \omega_{ik}\xi_{ijt} \le B^*_{jkt}, \quad j \in J', \ k \in K, \tag{23}$$

$$\sum_{j \in J'} \xi_{ijt} \le 1, \quad i \in I_t, \tag{24}$$

$$\xi_{ijt} \in \{0,1\}, \quad i \in I_t, \ j \in J', \ t \in T. \tag{25}$$

The expressions (22)-(25) represent the problem of linear integer programming with Boolean variables which has a considerably smaller dimension in comparison with the initial one. Indeed, if the admissible set of problems (13)-(19) is not an unfounded one then $|J'| < |J|$, since $J' = \{j \in J; \ \lambda_{ijt} > 0\}$ and J' is defined by the quantity of violated constraints in the course of distributions shaping. Besides, the variables of the choice of the development variant whose values are found at the stage of determination of admissible distribution (B^*_{jkt} - are the capacities corresponding to them) are absent in the problem (22)-(25). The transformations suggested allow not only to reduce the dimension but also to construct the simple algorithm for the derivation of the approximate solution of inverse problem. In combination with the branch - and - bound method this algorithm allows to estimate the intermediate solution and to obtain the optimal solution after a finite number of steps.

Let us consider the content of this algorithm. The criterion of assignment is a maximum of λ_{ijt} value and the condition is the realization of the corresponding constraints (23).

The solution found can be used as a lower estimate (of the record) in the scheme of branch - and - bound method.

To obtain the upper estimate one can solve the problem of linear programming (22)-(24) under condition of nonnegative variables. Taking into account the constraints of a "Knapsack" type (23) we suggest a simpler algorithm in comparison with the algorithm of the simplex - method. Its essence consists in successive assignment of units $i \in I_t$ according to the maximum criterion λ_{ijt} under condition of (23). If under the next assignment some constraint (23) is violated the admissible part of the order is assigned and the rest is considered as a new order which is distributed on general terms.

3. The Procedure of the subproblem solution

Let us write the Lagrange function for j-th subproblem.

$$\begin{aligned}
L_j &= \sum_t V_t[\sum_i c_{ijt}x_{ijt} + \sum_p [A_{pj} + g_{pj}(\Omega_{tj})]y_{pjt}] + \\
&\quad + \sum_t \sum_k [\sum_i \omega_{ik}x_{ijt} - \sum_{\tau=1}^t \sum_p B_{pjk}y_{pj\tau}]\beta_{kt}.
\end{aligned}$$

Following the transformations we shall obtain

$$L_j = \sum_t \sum_i [V_t c_{ijt} + \sum_k \beta_{kt}\omega_{ik}]x_{ijt} +$$
$$+ \sum_t \sum_p [V_t[A_{pj} + g_{pj}(\Omega_{jt})] - \sum_k [B_{pjk} \cdot \sum_{\tau=t}^{T} \beta_{k\tau}]]y_{pjt}. \tag{26}$$

Here, $\{\beta_{kt}\}$ are the Lagrange factors of constraints (10). The conditions (11)-(12) shape the area of function determination. Taking into consideration the constraints (9) the terms of the Lagrange function with the variables x_{ijt} will not be further considered.

Taking into account the above-mentioned remarks about the possibility of Lagrange function usage for the discrete programming we shall use the obtained expression for the choice of a branching variable in the branch - and - bound method.

In order to provide that the succession of variants is being exhausted in terms of increase of the objective function of subproblem we shall assume that in the expression (26) $B_{pjk} = \Omega_{kjt}$, if $\Omega_{kjt} > 0$. This will make the constraints (10) active without changing the area of optimal solutions for the j-th subproblem.

Then we shall find $\beta = \min_{\rho>0}(\max\{0, \beta_o + \rho\nabla_\beta L_j\})$, where $\nabla_\beta L_j$- is a vector - gradient of the L_j function in terms of Lagrange factors, ρ- is a value of step at which at least one coefficient in the expression (26) will be equal to 0 at y_{pjt}. Evidently, the corresponding variable would produce the maximum influence on the demand satisfaction within the time interval being considered.

Let us shape two sets on the basis of the opposite values of this variable that allow for the realization of the branching mechanism. The portion of variables fixed in this way will be called the partial solution. Then, if the value of the objective function of a subproblem will be taken as a estimate of a set chosen for the branching at the corresponding partial solution then the first admissible partial solution of one of the sets with the minimum estimation will be the optimal solution for j-th subproblem.

Thus, the algorithm will take the following form

1. Suppose, the index of the branching level is l=0 and the corresponding set is $G^l = 0$.

2. From the sets $G^o...G^l$ we choose the set G^q with the minimum estimate.

3. If the partial solution from the set G^q is admissible, then it is optimal. Therefore, the solution process is completed, otherwise, the iteration is continued.

4. In case of the order being not met

$$\Omega'_{kjt} = \max\{0, \Omega_{kjt} - \sum_{\tau=1}^{t} \sum_p B_{pjk} y_{pjt}\}$$

we shall find the variable for the branching y_{pjt}^* and form two sets $G^q = G^q \cup \{y_{pjt}^* = 0\}$, $G^{l+1} = G^q \cup \{y_{pjt}^* = 1\}$ taking into account the partial solution from the set G^q.

5. Let us assume $l := l + 1$ and pass on to the item 2.

4. The general algorithm for the problem solution

For the solution of the optimization problem we used the algorithm given below:

1. To obtain the initial admissible distribution of tasks $\xi = \{\xi_{ijt}\}$.

2. To solve the subproblems (8)-(12) and to determine the subgradient components $\delta W(\xi) = \{\delta W_j(\xi_j)\} = \{-u_{ijt}\}$.

3. To obtain a new distribution of tasks using the solution of the coordinating problem (13)-(19).

4. If a new distribution does not coincide with the preceding one we are to go over to item 2.

5. The solutions of subproblem corresponding to the minimum value of the functional $S = \sum_j W_j(\xi_j)$ is a solution of initial problem.

The obtained solution of the initial problem can be subjected to the estimation using the simulation models that allow for the estimation of such difficult - to - formalize index values as total losses due to the idling of task sources, waiting for the servicing or total losses due to the idling in consequence of service centers refusal, etc.

The balanced state between the minimum cost criteria and the minimum losses criteria can be reached through a successive use of the suggested general algorithm with the simultaneous change of the centralization level in the system. This change can be reached through the variation of the power increment variants within the framework of prescribed technical data of the equipment installed in the service centers.

5. Conclusion

The algorithms suggested have been realized in the form of program packages functioning under the medium VM-370 of the computer EC-1061 (an analog of IBM-PC desktop computer). The first variant was used for the determination of prospective development of helicopter maintenance system in Tyumen and the second variant was used for the determination of software - hardware of a distributed multiprocessor system exercising control over the technological equipment.

As an estimate of the efficiency of the algorithms being suggested we shall note that approximately 70% of the total time required for the solution of a problem is consumed by the coordinating problem (13)-919). And the estimation of "labor intensity" of the problem solution constitutes $O(m^2 n^2 t_m)$, where $m = \max_t\{|I_t|\}$, $n = |J|$, $t_m = |T|$.

References

1. G. J. Silverman, "Primal Decomposition of Mathematical Programs by Resource Allocation". *Techn. Memorandum 116, Operations Research Department, Case Western Reserve University, 1968.*

2. A. M. Geoffrion, "Primal Resource - Discrete Approaches for Optimizing Nonlinear Decomposable Systems". *Memorandum RM - 5829 - PR. The RAND Corporation, Santa Monica, Calif., 1968.*

Support Number and Allocation
for Optimum Structure

W. Gutkowski, J. Bauer, Z. Iwanow
Institute of Fundamental Technological Research,
Warsaw, Poland

1. Introduction

In structural optimization most of the attention has been paid so far to structure itself. However in many cases the type and number of supports are having an important influence on minimum cost of the total system - structure plus supports. To make the problem more clear let consider two different structures - bridge and guyed mast. The bridge symbolically presented on Fig.1 may be supported at seven structural joints 1,2,...,7. Depending on the conditions each of the seven supports may be of different price. One might be on rock other one on sand or bottom of river or sea. We have then to answer how to support our bridge to achieve the lowest cost, both of structural material and supports. With assumed configuration of structural nodes and members we may assume that the cost of assembling is constant and may be omitted in cost function.

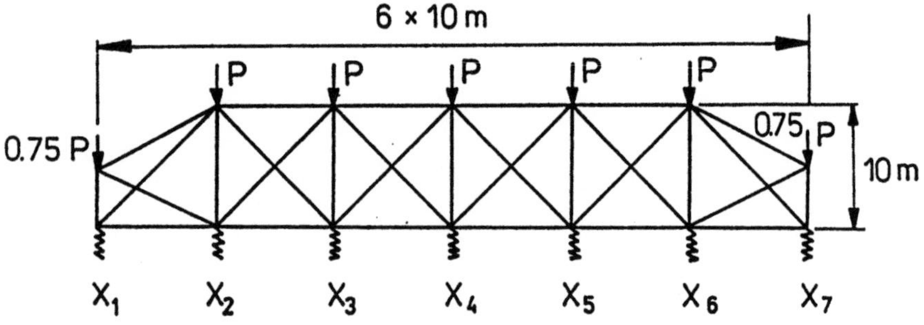

Fig. 1.

Height of a radio guyed mast, symbolically presented in Fig.2, is fixed as related to the radio wave. Other structural parameters may be seen as design variables. Among them is the number of supporting guys. Increasing guy number we are increasing also number of guy anchors and insulators. In the same time we are decreasing dimensions of mast members. This in turn causes decrease not only material volume but also external loading arising from wind pressure and icing.

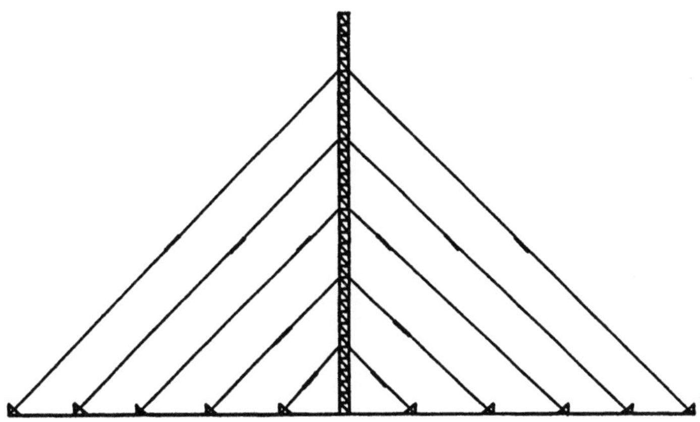

Fig. 2.

Two above examples are showing that the problem in question is multi-disciplinary one coupling together expenses on material and supports together with structural mechanics specifying forces, stresses and displacements. Other words in order to solve this kind of the problem, the relation between costs of particular structural members and structural supports with physical properties of these members and support must be known. In general these relations are nonlinear. In the paper with out loosing generality we are assuming these relations linear to make the reasoning more clear.

The idea of optimal allocation of structural supports is not a new one. However papers dealing with this problem [1], [2] were limited to relatively simple structures in form of a beam. So far to our knowledge no consideration dealing with an arbitrary structure and optimum number of support have been published.

Designing the structure from a finite set of available rolling profiles (list on catalogue of prefabricated steel cross section) our optimization problem becomes a discrete one. We also assume that in real structural design we may have to our disposal only a finite number of given kinds of supports with price of each of them related to maximal force it can bear.

There are several methods to solve a problem of discrete optimization. However only few of them efficient for relatively large problem. As large

170

problems we mean here order of magnitude $10^{10} - 10^{20}$ of possible combinations. One of the efficient method is controlled enumeration method which allows to solve such large problem within hours on PC [4]. The computing time depends of course on number of degrees of freedom of structural system. The method elaborated by authors in their earlier works was checked already on several examples [3] showing its practical effectiveness.

The computing time of discrete optimization problem is relatively long comparing with the some problem but with continuous variables. For this reason it is often better to apply as a first step of the solution the continuous approach, using it in discrete computation as lower bound value cost function.

The paper is illustrated with an example of bridge - like structure with three unknown structural cross section area and seven possible support positions.

2. Statement of the Problem

Consider structure of given configuration with f_o structural members or linking groups and i_o joints at which supports may be applied. Cross section area A^f of f-th structural member at cost of c^f per unit length may be chosen from a list of available rolled profiles defined by ordered pair

$$(A^f, c^f) \in [(A_1^f, c_1^f), ..., (A_{j_f}^f, c_{j_f}^f), ..., (A_{k_f}^f, c_{k_f}^f)] \tag{1}$$

with

$$f = 1, ..., f_o; \quad j_f = 1, ..., k_f$$

where

k_f - number of available profiles from which f-th
      structural element is chosen.

Support bearing maximum reaction R^i at cost C^i may be chosen from a list defined by ordered pair

$$(R^i, C^i) \in [(0,0), ..., (R_{l_i}^i, C_{l_i}^i), ..., (R_{m_i}^i, C_{m_i}^i)] \tag{2}$$

with

$$i = 1, ..., i_o; \quad l_i = 1, ..., m_i$$

where

m_i - number of available supports
in a list from which i-th support may be chosen.
We assume that $c_{j_f}^f$ and $C_{l_i}^i$ are ordered as follows

$$c_{j_f}^f < c_{(j+1)_f}^f; \quad C_{l_i}^i < C_{(l+1)_i}^i \tag{3}$$

The discussed problem may be stated as follows.
Find minimum of the cost C_d

$$C_d = \sum_{f=1}^{f_o} c^f l^f + \sum_{i=1}^{i_o} C_i \tag{4}$$

where l^f - length of f-th structural member.
The structure is subjected to following constraints

$$\mathbf{K(A)u - P} = 0 \tag{5}$$

$$\mathbf{u}^0 - \mathbf{u}_q \geq 0; \quad \sigma^o - \sigma \geq 0 \tag{6}$$

$$X_i \leq R^i_{j_i} \tag{7}$$

and sets of available profiles and supports defined by (1) and (2). X_i - stands for reaction at i-th support.

3. Methods of Solution

At least two methods may be applied to the solution of problem in question. The first one consists in finding an ordered sequence of values of the cost function (4). The combination of structural members and supports which gives the smallest value of the cost function (4) and at the same time fulfills all of the constraints (5)-(7) is the solution of our problem. It must be pointed out that there condition. The method of finding mentioned above sequence of values of the cost function was presented by Z. Iwanow [4] and will not be discussed in this paper. Bellow we will present an approach for case when continuous solution may be found and is considered to be the lower bound for the discrete one. Denoting by C_c^* minimum of the cost obtained by continuous approach and by C_d^* minimum discrete solution we have

$$C_c^* \leq C_d^*. \tag{8}$$

4. Controlled Enumeration Method with Given C_c^*

With above specified notations the cost function in this particular case the cost C_d of structure and its supports is given by (4). In order to present all possible values of C_d let consider a graph with tree structure constructed as follows (Fig.3)

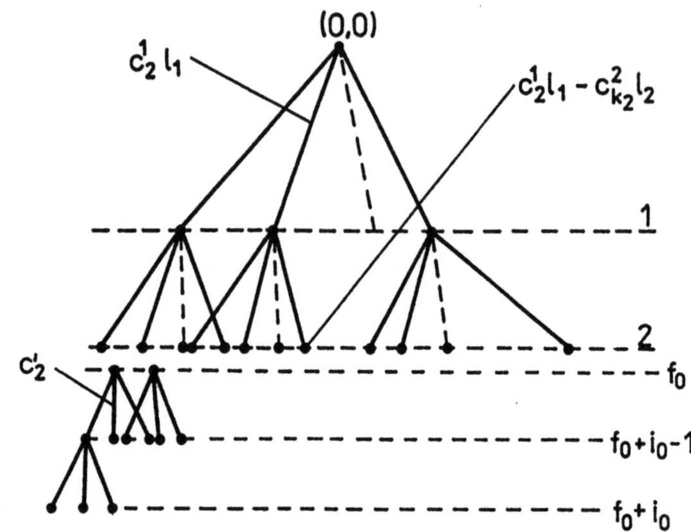

Fig. 3.

(i) Start with a root with pair of indices (0,0).

(ii) Subsequent vertices of the graph representing f_o it structural members are defined in such a way that the vertex with indices (f, j) becomes a parent of vertex $(f + 1, k)$ for $k = 1, ..., k_{f+1}$ up to $f = f_o$.

iii Subsequent vertices of the graph representing supports are defined in such a way that vertex with indices $(f_o + l, j)$ becomes a parent of vertex $(f_o + l + 1, k)$ for $k = 1, 2, ..., k_{i+1}$. The vertices with the first index $(f_o + i_o)$ don't have children.

(iv) Assign to the edge between the child with (f, j) and its parents the cost equal to $l_f c_{k_j}^f$ and to the edge between the child with $(f_o + l, j)$ the cost equal to $C_{m_j}^l$.

(v) Assign to the vertex with first index f (or $f_o + i$) the number equal to the sum of costs represented by all edges joining the vertex with the root $(0,0)$. All vertices with the first index f (or $f_o + i$) are said to belong to the layer f (or $f_o + i$).

Under above definitions vertices with the first index $f_o + i_o$ represent all possible combinations of values of the cost function C_d.

The number N of these combinations is equal to:

$$N = k_1 \cdot k_2 \cdot \ldots \cdot k_{f_o} \cdot m_1 \cdot m_2 \cdot \ldots \cdot m_{i_o} \tag{9}$$

According to (3) and (v) the extreme vertices belonging to the same layer of the graph (subgraph) represent the smallest and the largest cost from all possible costs included in the layer. This important property of the graph (subgraph) will be very useful in further considerations aimed to find the minimum of our discrete problem. Let now apply the described graph to an algorithm allowing to find minimum cost C_d^* of discussed structure together with its supports from lists of available structural members and supports. As it has been said the lower bound of our problem is the minimum cost of a structure for continuous design variables. First of all we then to find continuous relation between unit cost of structural members and their cross section areas, as well between cost of supports and their bearing reaction. This can be done by filleting curve method presented on Fig.4. This gives as needed relations which in our case, without loosing generality are assumed to be linear

$$c_{j_f} = r^f A_{j_f}^f, \quad C_{l_i}^i = z^i R_{l_i}^i \tag{10}$$

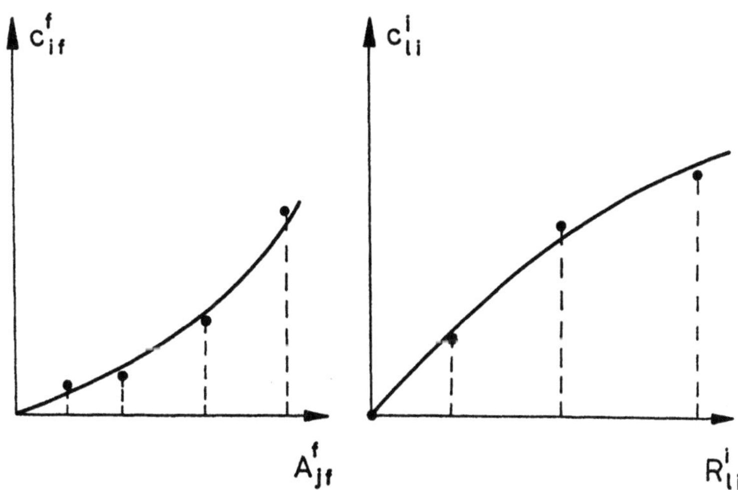

Fig. 4.

This is not aim of the present paper to discuss the minimum continuous solution C_c^*. It may be found applying algorithm presented in earlier Authors paper [3]. Other words for the purpose of presented paper we assume

174

that continuous minimum solution C_c^* is known. If so the minimum discrete solution C_d^* may be seen as minimum of all differences

$$\min(C_d - C_c^*) \tag{11}$$

for which all constraints are satisfied.

Below is the outline of an algorithm based on Iwanow algorithm [4] adapted to the purpose of discrete structural optimization.

STEP 1.
> Find the continuous solution for minimum cost C_c^* of the considered structure. The minimum cost C_c^* is the lower bound for discrete solution C_d^*.

STEP 2.
> Consider all subgraphs with roots $(1, k)$ $k = 1, 2, ..., k_1$ belonging to the layer 1. Extreme vertices in this layer $f_o + i_o$ represent the extreme values of the cost C_d and are as follows

$$C_{d1,i}^{\min} = l_1 c_{i_1}^1 + \sum_{f=2}^{f_o} l_f c_1^f \tag{12}$$

$$C_{d1,i}^{\max} = l_1 c_{i_1}^1 + \sum_{f=2}^{f_o} l_f c_{k_f}^f + \sum_{i=f_o+1}^{f_o+1} C_{m_i} \tag{13}$$

There are three possible relations between these numbers and continuous minimum solution C_c^*

$$C_i^* = C_{d_1,i}^{\min} \ (a); \quad C_{d_1,i}^{\min} \le C_c^* \le C_{d_1,i}^{\max} \ (b); \quad C_{d_1,i}^{\max} < C_c^* \ (c); \tag{14}$$

If (a) is true the combination of member sizes and supports represented by $C_{d_1,i}^{\min}$ and fulfilling constraints is solution of the problem if such a solution does not exist in subgraph $(1, j - 1)$.

If (b) is true go to STEP 3.

If (c) is true there is no solution of our problem among vertices of this layer.

STEP $3, 4, ..., f_o + i_o$.
> Apply the some procedure as in STEP 2 up to layer $f_o + i_o - 1$.

STEP $f_o + i_o + 1$.
> Enumerate vertices of $f_o + i_o$ layer which are children of vertices $(f_o + i_o - 1, j)$ and check following relation.

$$C_{d,f_o+i_o-1,j}^{\min} \le C_{d,f_o+i_o-1,j}^{\max} \tag{15}$$

Store vertices of each subgraph (f_o+i_o-1, j) for which $C_{d,f_o+i,j} - C_c^* = \min$.

STEP $f_o + i_o + 2$.

Successively verify if structures related to these combinations do not violate any of constraints. First structure fulfilling this condition is the solution of discrete minimum weight solution. If no one of these combinations fulfills constraint conditions go back to STEP 2 replacing C_c^* by the first stored combination of C_d [5].

5. Numerical Example

Let consider bridge - like structure of given configuration (Fig.1). There are seven lower structural nodes at which supports may be applied. The problem consists in finding number and allocations of supports giving minimum cost C_d of the total system equal to the cost of structural material together with the cost of supports:

$$C_d = \sum_{f=1}^{f_o} c^f l^f + \sum_{i}^{i_o} C_i$$

Design variables of the problem are as follows:

$A^1; A^2$ - cross section areas of upper and lower structural members respectively

A^3 - cross section area of vertical and diagonal members

R_i (i=1,2,...,7) reaction force which may be carried by i-th support.

The problem is subjected to following constraints.

Equilibrium equations

$$\mathbf{K}(\mathbf{A})\mathbf{u} - \mathbf{P} = 0$$

where $\mathbf{K}$, $\mathbf{u}$, $\mathbf{P}$ are stiffness matrix, nodal displacement vector and $\mathbf{P}$ external loading vector respectively.

Cross section are a A_j^f versus cost c_j^f of unit length of listed profiles and reaction capacity R_i versus cost of the i-th support C_i

$$C_j^f = 7500 A_j^f [zl/m] \text{ for all } f$$

$$C_{l_i}^i = z_i R_{l_i}^i$$

Inequality constraints are imposed on displacements $\mathbf{u}$ and stresses σ as follows

$$\mathbf{u}^0 - \mathbf{u} \geq 0 \quad \sigma^0 - \sigma \geq 0$$

where

$$u_i^0 = 0.1m; \qquad \sigma_f^0 = 294300.0 kPa$$

Beside that constraints are imposed on possible selection of design variables. Cross section areas of each structural member A^f may be chosen from following set

$$A^f := [0.001; 0.0014; 0.0018; 0.0028; 0.0033; 0.0040; 0.0047; 0.0063; 0.008; 0.01] m^2$$

176

Caring capacity R_i of each i-th support may be chosen from the set

$$R^i_{l_i} := [0; 490.0; 981.0; 1962.0]kN$$

with different price for each i-th support

$$z^i := [18.3490; 8.155; 12.232; 10.194; 8.155; 15.291; 12.232]zl/kN$$

The problem has been solved by algorithm [4] searching for increasing values of the cost function C. Each of these values is related to a certain combination of cross sections areas of structural members and supports. Each of these combinations in turn is checked if assumed constraints are not violated. The check is perform by ABAQUS program for FEM.

First value of the cost function containing support is the structure supported just at the middle of the structure. However this combination does not fulfil constraint imposed on the middle support which may carry up to $1962kN$ with the total reaction R_a equal to $2550.6kN$.

Finally after many iterations we come to the solution of minimum cost of our structure which is $C_d = 32050zl$. In this case the structure is supported at supports N_o 2 and 5 (Fig.5) and cross sections of structural members are $A_1 = 0.0033m^2$; $A_2 = 0.004m^2$; $A_3 = 0.0028m^2$.

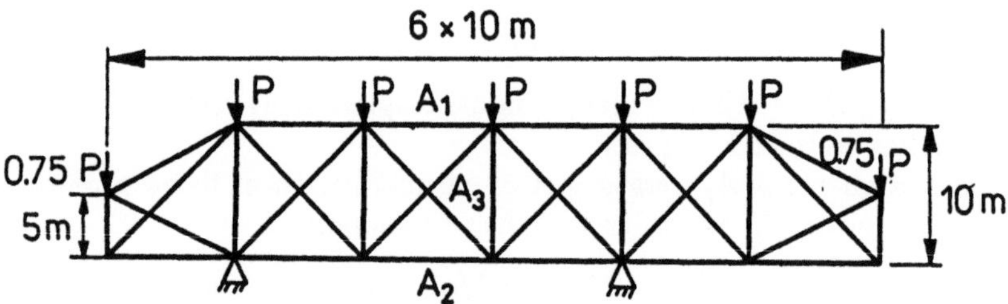

Fig. 5.

6. Conclusions

Presented approach allows to solve practically important problems of minimum cost for which not only the cost of structural material but also of structural supports must be taken into account. The presented method is based on controlled enumeration which helps to avoid the enumeration of all possible combinations arising in the problem. The consideration of all combinations makes the solutions of these type of problem practically impossible as their number may be of the order 10^{10} and more.

Acknowledgements. The paper was sponsored by the Polish Committee for Scientific Research grant *No* 3 0939 91 01. In numerical calculations ABAQUS program was applied.

References

[1.] Z. Mróz and G. I. N. Roavany, Optimal Design of Structures with Variable Support Conditions, *J. Opt. Theory and Appl.*, **15**, 1(1975), 85-101

[2.] M. B. Fuchs and M. A. Brull, A New Strain Energy Theorem and its Use in the Optimum Design of Continuous Beams, *Comp. Struct.*, **10** (1979), 647- 657

[3.] W. Gutkowski, J. Bauer and Z. Iwanow, Minimum Weight Design of Space Frames from a Catalogue [in:] Shells, Membranes and Space Frames, Proc. IASS Symp., Osaka, 1986, Ed. K. Heki, Elsevier, Amsterdam, 1986, Vol. 3, 229-236

[4.] Z. Iwanow, An Algorithm for Finding an Ordered Sequence of Values of a Discrete Linear Function, *Control and Cybernetics*, **19**, 3-4 (1990), 129-154

[5.] Z. Iwanow, The Method of Enumeration According to the Increasing Value of the Objective Function in the Optimization of Bar Structures, *Bull. Acad. Polon. Sci., Ser. Sci. Tech.*, **29**, 9-10 (1981), pp. 481-486

Sensor Allocation for State and Parameter Estimation of Distributed Systems

Józef Korbicz and Dariusz Uciński

Department of Robotics and Software Engineering
Technical University of Zielona Góra
PL-65-246 Zielona Góra, Poland

Abstract. This paper deals with the sensors allocation problem for the state estimation and parameter identification in non–linear stochastic distributed parameter systems described by parabolic partial differential equations. In order to pose the sensors location problem, a mathematical model of distributed systems is briefly considered. Then, a concise general review of several methods and approaches considered in the current literature for the state and parameter estimation is discussed. Some typical illustrative numerical example is presented in the final part of this survey type paper.

Keywords. Distributed parameter systems, sensors location, identification, state estimation.

1 Introduction

Many real physical systems are inherently distributed and their adequate mathematical description requires using the formalism of Partial Differential Equations (PDEs). Such systems are commonly called Distributed Parameter Systems (DPSs) (see e.g. Omatu and Seinfeld, 1989; Korbicz and Zgurowski, 1991), and typical examples are many industrial processes where heat and mass transfer and chemical reactions are considered. They also include e.g. structural analysis and design where vibrations and dynamic behavior are central, acoustics problems, underground water and oil exploration, air and water pollution propagation and nuclear energetics (see e.g. Tzafestas, 1982; El Jai and Amouroux, 1989).

A fundamental problem in distributed systems control is measurement. In many industrial processes the nature of the state variables does not allow much flexibility as to which states can be measured. For variables that can be measured on–line, it is usually possible to make the measurements continuously in time. However, since it is generally impossible to measure process states over the entire spatial domain of the systems, the spatial location of sensors and the information content of the resulting signals with respect to the distributed state and PDE model become the primary questions.

The measurement strategies which have been considered in the literature are as follows:

- stationary location of a given number of sensors (e.g. Kubrusly and Malebranche, 1985),

- movable sensors (e.g. Carotenuto *et al.*, 1987),

- scanning, i.e. using only some part of all the stationary located sensors at the moments when the measurements can be taken (e.g. Nakano and Sagara, 1988; Korbicz, 1991).

As a matter of fact, every real transducer averages over some portion of the spatial domain. In most applications, however, this averaging is approximated by considering that pointwise measurements are available at a number of spatial locations. In other case, the problem of finding the optimal geometry of the sensor support has been posed by El Jai and Pritchard (1988), and El Jai (1991).

The above mentioned techniques involve the following problems, respectively:

- how to choose the minimal number of the sensors and their location,

- how to choose their optimal trajectories and velocities,

- how to choose optimal location of the sensors at given moments.

Furthermore, from the engineering point of view the sensors must be located so that the configuration fulfils the requirements of system identification, state estimation and optimal control, and in addition the technical constraints.

A vast amount of literature has been produced dealing with the optimal sensor location problem for DPS, especially in the context of state estimation (see e.g. Kubrusly and Malebranche (1985) for the survey of the state of the art in the mid 1980s or the book by El Jai and Pritchard (1988)). There have been much fewer works on this subject related to parameter identification. This is mainly because the direct extension of the appropriate results from state estimation is not straightforward and has not been pursued. The problems are essentially different since in the latter case the state depends strongly non–linearly on unknown parameters.

In this paper the sensor location problem for the state estimation and identification in non–linear stochastic distributed systems is discussed. In order to pose the sensor location problem, at the beginning the distributed systems described by parabolic partial differential equations are considered. Then, a concise general review of several methods and approaches known in the current literature for state estimation and identification is presented. Special attention is paid to suboptimal solutions which can be obtained for most of the practical problems, e.g. air pollution monitoring, control of flexible space structures *et al.* An illustrative numerical example is shown in the last part of this survey type paper.

180

2 System Description

Let Ω be a simply connected, open set in $\mathbb{R}^r$ with smooth boundary $\partial\Omega$, $\overline{\Omega} = \Omega \cup \partial\Omega$. Points of Ω (spatial coordinate vectors) will be denoted by $x = (x_1, x_2, ..., x_r)$, and the time by t, $t \in (0, t_f]$. The system is assumed to be described by the non-linear stochastic partial differential equation (Korbicz *et al.*, 1988)

$$\frac{\partial y(x,t)}{\partial t} = \mathcal{N}_x(y, x, t, \theta) + B(x,t)w(x,t), \quad x \in \Omega, \ t \in (0, t_f] \quad (1)$$

where $y(x,t)$ denotes the n-dimensional state vector, $B(x,t)$ is a known $(n \times q)$-matrix function, $\mathcal{N}_x(\cdot)$ is a non-linear spatial n-dimensional differential operator vector and $w(x,t)$ is the q-dimensional vector of the Gaussian stochastic processes with zero mean and the covariance matrix $Q(x, s, t)$, $x, s \in \Omega$, and $\theta(x,t)$ is the s-dimensional unknown parameter vector.

The initial and boundary conditions for (1) are given, respectively, by

$$y(x, t)\Big|_{t=0} = y_0(x), \quad x \in \Omega \quad (2)$$

$$\mathcal{N}_{bx}(y, x, t, \theta_b) = 0, \quad x \in \partial\Omega, \ t \in (0, t_f] \quad (3)$$

where $\mathcal{N}_{bx}(\cdot)$ is a non-linear spatial n'-dimensional differential operator vector defined on $\partial\Omega$, n' is dimension defined by the boundary condition type, and $y_0(x)$ is the n-dimensional vector of the Gaussian stochastic processes with known statistics given by $\bar{y}_0(x)$, $P_0(x, s)$, and $\theta_b(t)$ is the s'-dimensional parameter vector.

Furthermore, the system described by (1)–(3) is assumed to be well-posed in the sense of Hadamard, i.e. the solution exists uniquely and depends continuously both on the initial (2) and boundary conditions (3).

As regards observations, we assume that measurements of the state $y(x, t)$ are made continuously in time at N fixed points x^j, $j = 1, ..., N$ of Ω and the observation system is non-inertial, i.e.

$$z(t) = H(t)y_N(t) + v(t) \quad (4)$$

where $z(t) = \mathrm{col}\{z(x^j, t), \ j = 1, 2, ..., N\}$ is the $N \cdot m$-dimensional measurement vector, $y_N(t) = \mathrm{col}\{y(x^j, t), \ j = 1, 2, ..., N\}$ is an $N \cdot n$-dimensional state vector at the measurement points, $H(t) = \mathrm{diag}\{H_j(t) = H(x^j, t), \ j = 1, 2, ..., N\}$ is a known $(N \cdot m \times N \cdot n)$ block-diagonal matrix and $v(t) = \mathrm{col}\{v_j(t) = v(x^j, t), \ j = 1, 2, ..., N\}$ is the $N \cdot m$-dimensional vector of white Gaussian processes with zero mean and the covariance matrix $R(t)$.

The white Gaussian processes $w(x, t)$ and $v(t)$ are assumed to be statistically independent of one another and also independent of the stochastic initial condition $y_0(x)$.

Taking into account the foregoing, we wish to obtain estimates $\widehat{y}(x, t; \widehat{\theta})$ of the state $y(x, t; \theta)$ based on all the past measurements $z(t')$, $t' \in (0, t)$. The estimate $\widehat{y}(x, t; \widehat{\theta})$ which minimizes the variance $E\|y(x, t; \theta) - \widehat{y}(x, t; \widehat{\theta})\|^2$ is called the optimal one, where $\| \cdot \|$ denotes the Euclidean norm. It is a strongly non–linear estimation problem which can be solved only by using some approximation methods (Tzafestas, 1982; Omatu and Seinfeld, 1989; Korbicz and Zgurovsky, 1991). On the other hand, these solutions depend on the measurement vector $z(t)$ and therefore we should look for the points $x^j \in \Omega$, $j = 1, 2, ..., N$ of sensors location which are optimal in some sense. Below, the sensors location problem for identification and state estimation are considered separately.

3 Sensors Location for State Estimation

Up to date the optimal sensor location problem has been solved by many authors using different methods and approaches. Basic results have been obtained for the linear filtering theory of DPS (see Tzafestas, 1982; Omatu and Seinfeld, 1989). A common feature of the methods applied in these books is the necessity to solve the non–linear Riccati–type equations for the covariance matrix of the estimation error resulting from the Kalman filter. A great computation burden of these algorithms has stimulated many authors to look for suboptimal solutions of the sensors location problem. A computationally efficient procedures have been proposed by Kumar and Seinfeld (1978), Nakamori *et al.* (1980) which minimize an upper bound of the estimation error variance. The allocation algorithm which incorporates the recursive selection of the sensors location into the equation which defines the evolution of the covariance has been considered by Carotenuto *et al.* (1990). Contrary to the methods mentioned above, Magami's *et al.* (1990) approach reduces the problem of sensor positioning to a solution of a non–linear mathematical programming problem. This method was applied to the sensor positioning of a large–order flexible space structure.

Although the optimal measurement problem for movable sensors is very attractive from the viewpoint of the degree of optimality, only a few authors have examined this topic. Carotenuto *et al.* (1987) have minimized their complex optimality criterion and the resulting mathematical statement of moving sensors problem has been transformed to an optimal control problem for a lumped–distributed parameter system. An alternative approach to moving sensors is to assume that several sensors are placed in predetermined locations, and the state is measured from one of them at a time (Nakano and Sagara, 1988; Korbicz, 1991). It is called the scanning observation problem.

The majority of known research has been devoted to the optimal measurement problem for the linear DPS and with less work on the non–linear systems. Using the suboptimal Kalman filtering algorithm for DPS, a computationally efficient method was proposed by Korbicz *et al.* (1988). Below, main features of this method are discussed.

3.1 Sensors Location Problem

It is necessary to define a minimal number of sensors N and fixed points of their location $x^{\mathrm{opt}} = \left[x^1 x^2 ... x^N\right]$ at domain $\overline{\Omega} = \Omega \cup \partial\Omega$ which minimize the cost function

$$J(x) = \int_0^{t_f} \int_\Omega \int_\Omega \mathrm{tr}\left[\widehat{P}(x, s, \tau)\right] \, \mathrm{d}x \, \mathrm{d}s \, \mathrm{d}\tau \tag{5}$$

where $\mathrm{tr}\left[\cdot\right]$ denotes the trace of matrix $\left[\cdot\right]$, $\widehat{P}(x, s, \tau)$ is the covariance matrix function estimate, with respect to restrictions given by the suboptimal Kalman filter (Korbicz $et\ al.$, 1988)

$$\frac{\partial\widehat{y}(x, t)}{\partial t} = \mathcal{N}_x(\widehat{y}, x, t) + \widehat{P}_N(x, t) H^T(t) R^{-1}\left[z(t) - H(t)\widehat{y}_N(t)\right] \tag{6}$$

$$\frac{\partial\delta y(x, t)}{\partial t} = (\mathcal{N}_x)_{\widehat{y}} \, \delta y(x, t) - \widehat{P}_N(x, t) H^T(t) R^{-1}(t) \tag{7}$$

$$\frac{\partial\widehat{P}(x, s, t)}{\partial t} = -\frac{1}{T_t}\widehat{P}(x, s, t) + \frac{1}{T_t T_x T_s} \int_{\Omega_x} \int_{\Omega_s} \delta y(\gamma, t) \, \delta y^T(\rho, t) \, \mathrm{d}\rho \, \mathrm{d}\gamma \tag{8}$$

with known boundary and initial conditions. Above $\widehat{y}(x, t)$ denotes the n–dimensional state estimation vector; $\delta y(x, t) = y(x, t) - \widehat{y}(x, t)$ is the filtering error vector; $(\mathcal{N}_x)_{\widehat{y}} = \partial\mathcal{N}_x/\partial y\big|_{y=\widehat{y}}$ denotes the Jacobian matrix; T_t, T_x and T_s are the chosen constant coefficients; $\widehat{P}(x, s, t)$ is the $(n \times n)$ filtering error covariance matrix and $\widehat{P}_N(x, t)$ denotes the matrix at the measurement points. It should be noticed that filter equations (6)–(8) are obtained for DPS model (1)–(4) assuming that parameter vector θ is known.

Solution of optimization problem (5)–(8) can be done by its transforming to the bang–bang optimal control problem. After that, one can obtain the computational effective recursive algorithm of the sensors location (Korbicz $et\ al.$, 1988). The computational efficiency of this algorithm is defined by using the suboptimal filtering algorithm (6)–(8) which does not require solution of the Riccati–type equations for the filter error covariance. Up to now there have been many practical applications of this recursive algorithm of sensors location (see e.g. the monitoring stations location problem for air pollution processes (Korbicz and Gawłowicz, 1989), the minimal number choice of drilling rigs at the oil and gas deposit (Zgurovsky $et\ al.$, 1985).

4 Optimal Sensors Allocation for Parameter Identification

Unlike the optimal sensors location problem for state estimation discussed above, only a few papers have appeared on this problem for DPS identification. This is caused by serious difficulties related to the non–linear dependence between the system state and the unknown parameters to be identified.

Nevertheless, the existing methods can be divided into three main groups, namely:

- methods leading to state estimation,
- methods using theory of optimum experimental design,
- methods leading to random fields analysis.

The methods belonging to the first group consist in transforming the problem into the state estimation one (by augmentation of the state space), and then using one of the methods considered in Section 3. However, since the state and parameter estimation are then carried out simultaneously, the whole problem becomes strongly non–linear. To overcome this difficulty, a sequence of linearizations at suitable trajectories is performed by Malebranche (1988) and a special suboptimal filtering algorithm is used by Korbicz *et al.*, (1988).

The second class of the methods is closely related to classical optimum experimental design. The optimal sensors location criteria are essentially the same: maximization of a scalar measure of the Fisher information matrix associated with the parameters to be identified. It is worth noting that the inverse of this information matrix is a solution of the Riccati equation, related to the state estimation problem, for the case of no input noise. Two simple examples are given by Quereshi (1980). In (Rafajłowicz, 1981) the information matrix was associated to the system eigenvalues rather than to the system parameters. Conditions for optimality of the experiment design were derived after some simplifications (among others the infinite observation time and restricted form of the inputs were assumed). The sensor placement problem from the standpoint of a structural dynamicist who must use the data collected from the sensors to validate a large space structures finite element model was studied by Kammer (1990). There, the sensors configuration tends to maximize the determinant of the appropriate Fisher information matrix.

The third group of the methods is very interesting for applications. This approach is based on random fields theory. Since DPS are described by PDEs, direct use of that theory is impossible, and therefore this decription should be replaced by the characteristics of the random field, e.g. the mean and the covariance function. Such method for some mechanical system subjected to the action of a random load has been considered by Kazimierczyk (1989). For optimum sensors placement in random fields analysis the theory of optimum experimental design turns out to be very effective (Brimkulov *et al.*, 1986).

All the above–mentioned techniques concerning parameter identification are limited to the case of stationary sensors, but the optimal measurement problem for spatially movable sensors has been studied, as well. Rafajłowicz's (1986) approach is closely related to the methods of optimum experimental design for lumped–parameter system identification and it is based on looking for the optimal time–dependent measure rather than for the trajectories themselves. Nevertheless, their use is restricted to relatively simple systems.

184

On the other hand, Uciński *et al.* (1993) have presented an algorithm for the choice of movable sensor trajectories maximizing the trace of the Fisher information matrix as a measure of the parameter identification accuracy for two–dimensional distributed systems. In the next section the main features of this approach are considered.

4.1 Sensors Allocation Problem

Assume that system description (1)–(4) is given in the form of deterministic PDE's with noisy observations, i.e. only $v(t)$ is stochastic, whereas $y_0(x)$ and $w(x,t)$ are completely known. The measurements are usually of the form

$$z_j(t) = h_j(y(x^j,t),t) + v(x^j,t), \quad t \in [0,t_f], \; j = 1,...,N \tag{9}$$

where spatially uncorrelated white noise is considered with covariance $S\delta_{jk}\delta(t-\tau)$, where δ_{jk} and $\delta(t-\tau)$ denote the Kronecker and Dirac delta functions, respectively.

The Fisher information matrix M for the unknown parameter vector θ to be identified is then given by Quereshi *et al.* (1980)

$$M = \frac{1}{t_f} \sum_{j=1}^N \int_0^{t_f} \frac{\partial y^T(x^j,t)}{\partial \theta} \frac{\partial h_j^T(y(x^j,t),t)}{\partial y} \times$$

$$S^{-1} \frac{\partial h_j(y(x^j,t),t)}{\partial y} \frac{\partial y(x^j,t)}{\partial \theta} \, \mathrm{d}t \tag{10}$$

It is worth noting that the inverse of M is the solution of the Riccati equation, related to the state estimation problem, for the case of no input noise.

For fixed inputs, M is only a function of the sensor characteristics (i.e. their structure) and the sensor location. In practical cases the former ones are fixed for the given type of sensors. This fact implies that the Fisher matrix depends only on the sensors locations and these can be optimized by choosing x^j, $j = 1,..,N$ so as to maximize some scalar measure of the information matrix. Various choices exist for such scalar function including the following (Yermakov and Zhiglavsky, 1987; Zarrop and Goodwin, 1975): i) $-\mathrm{tr}\left[LM^{-1}\right]$, ii) $\det[M]$, and iii) $\mathrm{tr}[LM]$, where L is a positive definite weighting matrix. The advantage of using choice (iii) above is that it leads to simple design algorithms (Kalaba and Spingarn, 1982). Its maximixation increases sensitivity of the outputs with respect to the unknown parameters. The criterion ii) seems to be preferred by most researchers (this leads to the well–known D–optimal designs). It is noteworthy to add that in the case of one unknown parameter all the above criteria are equivalent. In order to determine the elements of matrix M, it is necessary to know how to compute the measurements sensitivities (or state sensitivities) with respect to the unknown parameters. Some simple procedure is described by Uciński, (1992).

There are two main difficulties complicating the problem of optimizing the identification accuracy by the appropriate sensors location. First, in most cases the solution depends on the unknown parameters (Quereshi *et al.*, 1980) which only should be identified. One can doubt that the optimization of the sensors location makes sense in such a case. The implication is simply that the design should be based on nominal parameters values which can be obtained by a preliminary experiment or by physical analysis. Another problem is that under the optimal experimental conditions one can observe sensors clusterization, i.e. sensors are assembled at some points of the spatial domain. But this occurence is only apparently strange. It results from the assumption that measurement from one sensor have no influence on measurements from the other one (the measurement noise is Gaussian, spatial uncorrelated and white) and that spatial dimensions of the sensors can be neglected.

5 Example

In order to illustrate the foregoing, let us consider a simple two–dimensional heat conduction process described by the equation

$$\frac{\partial y(x_1, x_2, t)}{\partial t} = \mu \left(\frac{\partial^2 y(x_1, x_2, t)}{\partial x_1^2} + \frac{\partial^2 y(x_1, x_2, t)}{\partial x_2^2} \right) \tag{11}$$

$$(x_1, x_2) \in \Omega = (0, 2) \times (0, 2), \ t \in (0, 0.5]$$

I.C. $y(x_1, x_2, 0) = 0, \qquad (x_1, x_2) \in \Omega$

B.C. $y(x_1, x_2, t) = \begin{cases} 100, & (x_1, x_2) \in (0, 2) \times \{0\}, \ t \in (0, 0.5] \\ 0, & (x_1, x_2) \in \partial\Omega \setminus (0, 2) \times \{0\}, \ t \in (0, 0.5] \end{cases}$

This case may correspond to a thin square copper plate, one side of which is heated up (e.g. with the use of an infra–red heater) and the other ones are insulated.

The measurements made by a thermocouple can be considered to be pointwise and are given by

$$z(x_1^1, x_2^1, t) = y(x_1^1, x_2^1, t) + v(x_1^1, x_2^1, t) \tag{12}$$

where (x_1^1, x_2^1) denotes the location of the thermocouple. We assume that the measurement noise v is Gaussian, spatial uncorrelated and white with known characteristics.

The objective of the measurements is to estimate the unknown parameter μ (assumed to be constant) from the output data as accurately as possible over the period of observation.

For the D-optimality criterion, this optimum measurement location problem reduces to determining the location (x_1^1, x_2^1) such that $\det M$ is maximized. Figures 1a and 1b show the appropriate surface plot and contour map obtained with the use of an algorithm described by Uciński (1992) for $\mu = 1.0$. The results suggest that the best choice is to locate the sensor at the point $(1.0, 0.5)$ which fully agrees with our intuition.

6 Conclusions

From engineering point of view it is clear that sensor placement is an integral part of the control design, particulary in the control of distributed parameter systems, e.g. the flexible structures, the air pollution processes, the oil and gass production from deposit. On the other hand, the engineering judgement and trial and error analysis are quite often used to determine the optimal locations of sensors. It seems to us that two main reasons can explain why strongly formal methods do not work in engineer's practice. First, only relatively simple engineering problems can be solved using the linear approximation of the practical problem. Second, the numerical complexity of most of the sensor location algorithms does not encourage engineers to apply them in their practice.

This paper briefly reviewed the state of the art in the sensors location problem for state and/or parameter estimation in distributed systems. Two methods presented in more detail (see Secs. 3.1 and 4.1) showed that their application in practice can be simpler (methods of the sensors location which do not need solution of the Riccati–type equations) and, what is very important, both the sensors location for state estimation (Sec. 3.1), and for identification (Sec. 4.1) can be applied to solve complex non–linear problems.

Furthermore, as it was emphasized by Kubursly and Malebranche (1985), more attention should be paid to the problem of sensor allocation for parameter idenfitication of DPS. The use of the existing methods is restricted because they usually involve computational and/or realizing difficulties. Little literature has been written about the results regarding two– or three–dimensional spatial domains and spatially–varying parameters. Thus, some generalizations are still expected in this connection. Apart from that, some strategies in the case of arbitrary (not only Gaussian) probability distributions for the measurement noise are still lacking. As already remarked, most of the contributions deal with the choice of stationary sensors positions. On the other hand, the optimal measurement problem for spatially movable sensors seems to be very attractive from the viewpoint of the degree of optimality and should receive more attention. Finally, more efforts towards the joint optimization of the input signals and the sensors location should be attempted.

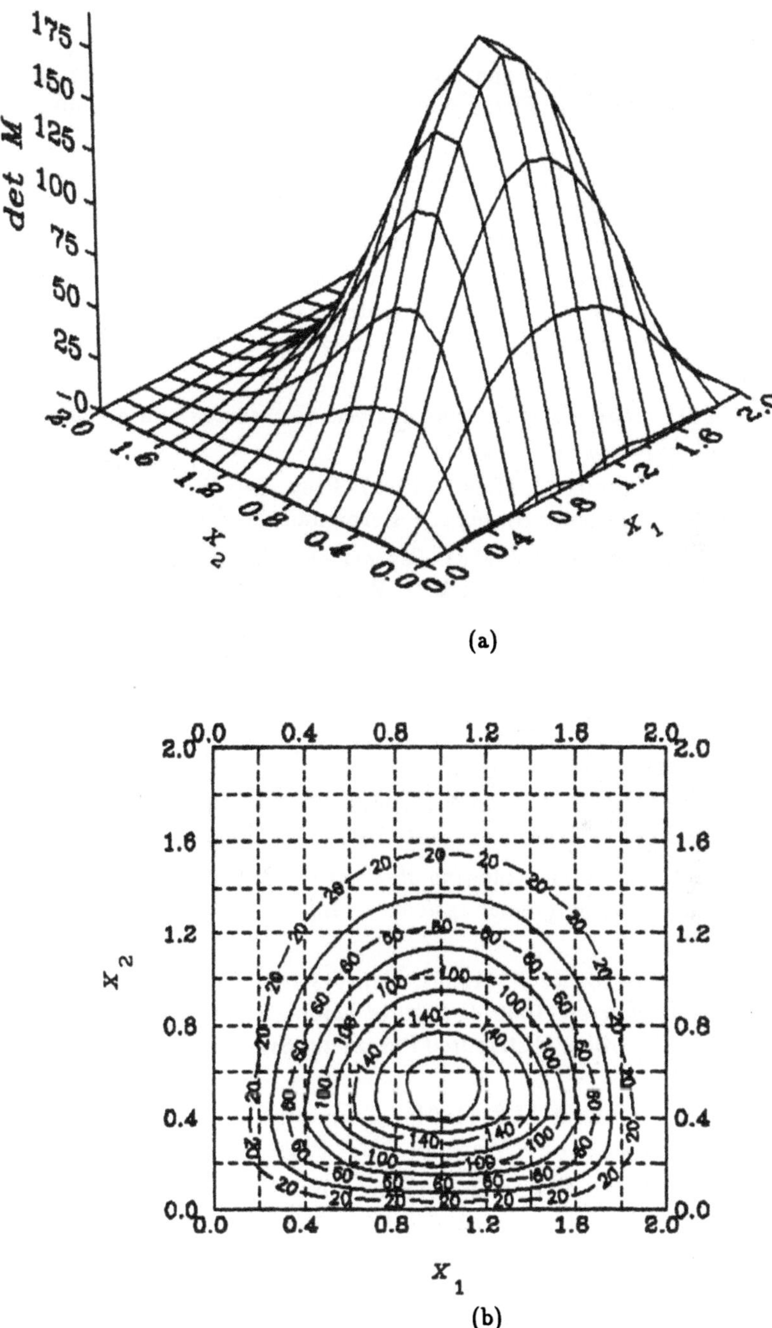

(a)

(b)

Fig. 1. Variation of $\det M$ with the sensor location:
(a) the surface plot   (b) the contour map.

References

Brimkulov U.N., Krug T.K. and Savanov V.L. (1986): *Experiment Design in Research of Random Fields and Processes.* — Moscow: Nauka, (in Russian).

Carotenuto L., Muraca P. and Raiconi G. (1987): *Optimal location of a moving sensor for the estimation of a distributed-parameter process.* — Int. J. Control, v.46, No.5, pp.1671–1688.

Carotenuto L., Muraca P., Pugliese P. and Raiconi G. (1990): *Recursive sensor allocation for a class of distributed parameter systems.* — Proc. IASTED Int. Symp. *Modelling, Simulation and Optimization,* Montreal, Canada.

El Jai A. (1991): *Distributed systems analysis via sensors and actuators.* — Sensors and Actuators, v.29, pp.1–11.

El Jai A. and Amouroux M. (Eds.) (1989): Prep. 5th IFAC Symp. *Control of Distributed Parameter Systems.* – Perpignan, France.

El Jai A. and Pritchard A.J. (1988): *Sensors and Controls in the Analysis of Distributed Systems.* — New York: J. Wiley.

Kalaba R. and Spingarn K. (1982): *Control, Identification and Input Optimization.* — New York: Plenum Press.

Kammer D.C. (1990): *Sensor placement for on-orbit modal identification and correlation of large space structures.* — Proc. *American Control Conf.,* San Diego, California, USA, v.3, pp.2984–2990.

Kazimierczyk P. (1989): *Optimal Experiment Design; Vibrating Beam Under Random Loading.* — Warszawa: Inst. PPT PAN Press, (in Polish).

Korbicz J. (1991): *Discrete-scanning observation problem for stochastic non-linear discrete-time parameter system.* — Int. J. Systems Sci., v.22, No.9, pp.1647–1662.

Korbicz J. and Gawłowicz P. (1989): *Sensors location problem for stochastic non-linear discrete-time distributed parameter systems.* — Prep. 5th IFAC Symp. *Control of Distributed* Parameter Systems, Perpignan, France, pp.421–426.

Korbicz J. and Zgurowsky M.Z. (1991): *Estimation and Control of Distributed Parameter Systems.* — Warsaw: PWN Press, (in Polish).

Korbicz J., Zgurovsky M.Z. and Novikov A.N. (1988): *Suboptimal sensors location in the state estimation problem for stochastic non-linear distributed parameter systems.* — Int. J. Systems Sci., v.19, No.9, pp.1871–1882.

Kubrusly C.S. and Malebranche H. (1985): *Sensors and controllers location in distributed systems: a survey.* — Automatica, v.21, No.2, pp.117–128.

Kumar S. and Seinfeld J.H. (1978): *Optimal location of measurements for distributed parameter estimation.* — IEEE Trans. Automat. Control, v.AC-23, No.4, pp.690–698.

Magami P.G. and Joshi S.M. (1990): *Sensor/actuators placement for flexible space structures.* — Proc. American Control Conf., San Diego, California, USA, v.2, pp.1941–1047

Malebranche H. (1988): *Simultaneous state and parameter estimation and location of sensors for distributed systems.* — Int. J. Systems Sci., v.19, No.8, pp.1387–1405.

Nakamori Y., Miyamoto S., Ikeda S. and Sawaragi Y. (1980): *Measurement optimization with sensitivity criteria for distributed parameter systems.* — IEEE Trans. on Autom. Control, v.AC–25, No.5, pp.889–900.

Nakano K. and Sagara S. (1988): *Optimal scanning measurement problem for a stochastic distributed parameter system.* — Int. J. Syst. Sci., v.19, No.7, pp.1429–1445.

Omatu S. and Seinfeld J.H. (1989): *Distributed Parameter Systems. Theory and Applications.* — Oxford: Oxford University Press.

Quereshi Z.H., Ng T.S. and Goodwin G.C. (1980): *Optimum experimental design for identification of distributed parameter systems.* — Int. J. Control, v.31, No.1, pp.21–29.

Rafajlowicz E. (1981): *Design of experiments for eigenvalue identification in distributed parameter systems.* — Int. J. Control, v.34, pp.1079–1094.

Rafajlowicz E. (1986): *Optimum choice of moving sensor trajectories for distributed parameter system identification.* — Int. J. Control, v.43, No.5, pp.1441–1451.

Tzafestas S.G. (Ed.)(1982): *Distributed Parameter Control Sytems. Theory and Application.* — London: Pergamon Press.

Uciński D. (1992): *Optimal sensors location for parameter identification of distributed systems.* — Appl. Math. and Comp. Sci., v.2, No.1, pp.119–134.

Uciński D., Korbicz J. and Zaremba M. (1993): *On optimization of sensors motions in parameter identification of two-dimensional distributed systems.* — Proc. 2nd European Control Conference, Groningen, The Netherlands, v.3, pp.1359–1364.

Yermakov S.M. and Zhiglavsky A.A. (1987): *Mathematical Theory of Optimum Experiments.* — Moscow: Nauka, (in Russian).

Zarrop M.B. and Goodwin G.C. (1975): *Comments on "Optimal inputs for system identification".* — IEEE Trans. Automat. Control, No.2, pp.299–300.

Zgurovsky M.Z., Korbicz J. and Novikov A.N. (1985): *Numerical calculation of parameters and states of saturated porous medium and optimal sensors location.* — Chemical Technology, No.2, pp.50–55 (in Russian).

Optimal Location of Piezoelectric Actuators

Jan Holnicki-Szulc[1], Francisco López-Almansa[2] and Andrzej Maćkiewicz[3]

[1] Institute of Fundamental Technological Research
Świętokrzyska 21, 00-049 Warsaw, Poland
(Visiting Professor at the Universidade da Beira Interior, Covilhã, Portugal)

[2] Universitat Politécnica de Catalunya, Barcelona, Spain

[3] Technical University of Poznań, Institute of Mathematics,
Piotrowo 3a, Poznań, Poland
(Visiting Professor at the Universidade da Beira Interior, Covilhã, Portugal)

Abstract. A new model of flexible plate (equipped with piezoelectric wafers) applicable for design and simulation of active control process is demonstrated. In particular, a concept of continuously distributed 'active distortions' describing interactions between piezoelectric wafers and the structure itself allows us to formulate properly the problem of optimal distribution for these control devices. The standard way to describe the interaction between piezoelectric devices and the controlled structure by means of concentrated forces applicable at the edge of actuator is not suitable to discuss the optimal location problem. Restricting control action to a chosen set of independently modificable eigenmodes of vibration and applying the criterion of maximization of available substitutive control action the discrete optimization problem can be finally formulated. A numerical procedure for solving this problem and corresponding testing example is presented.

Keywords. Active control, Piezoelectric actuators, Discrete optimization, Integer programming, Branch-and-bound metchod.

1 Introduction

The main problem in design of active control of flexible structures is optimal location of sensors and actuators (cf. Ref. 1, 2,4, 6, 9, 10) as well as robost control strategy. The robostnes of the control process can be strongly affected if the position of actuators are not defined precisely. The application of piezoelectric wafers to control vibration of flexible plates has been discussed in Ref.3. The mechanical model considered there allows us to simulate the behavior of the plate with two sticked p-e wafers (on the both, upper and bottom surfaces). However, it does not supply us with a tool necessary for design of optimal location for these p-e actuators.

The aim of this paper is to present a model of the plate equipped with the p-e devices applicable to design and simulation of the active control process. In particular, a new method for optimal design of locations for actuators (controlling chosen eigenmodes) based on the discrete optimization technique is applied. The presented method explores the concept of 'active distortions' generated by actuators and therefore, the main effort of this paper is dedicated to determination of these active distortions (distributed continuously along the plate) simulating the influence of the p-e wafers on the structure. The presented method, tested in this paper on the one dimensional example of beam vibration will be discussed more generally in application to active damping of plate vibration in Ref.7.

2 The 'Active Distortion' Model of Plate Equipped with Piezoelectric Wafers

Let us consider a thin plate equipped with two p-e wafers sticked to the upper as well as to the bottom surface of the plate (cf. Fig.1). Assuming the Kirchoff-Love kinematic hypothesis according to the components of the displacement field:

$$\begin{aligned}
u_3 &= u_3(x_1, x_2) \\
u_i &= -x_3 u_{3,i}
\end{aligned} \tag{1}$$

where i=1,2, the equilibrium equation can be written in the following form

$$m_{ij,ij} = \rho \, \ddot{u}_3 - f \tag{2}$$

where ρ denotes the mass density per unit area and m_{ij} denotes the components of the bending moment tensor related to the planar stresses as follows:

$$m_{ij} = \int_{-h}^{h} x_3 \sigma_{ij} dx_3 \quad on \quad \varpi - \varpi_p \tag{3}$$

$$m_{ij} = \int_{-h-\alpha}^{h+\alpha} x_3 \sigma_{ij} dx_3 \quad on \quad \varpi_p \tag{4}$$

192

Restricting our considerations to the case $\sigma_{ij}(x_3) = -\sigma_{ij}(-x_3)$ and therefore, constraining the discussion to the only pure bending actions of the p-e
devices the relations (4) can be written in the form

$$m_{ij} = \int_{-h}^{h} x_3\sigma_{ij}\,dx_3 + 2\int_{h}^{h+\alpha} x_3\sigma_{ij}\,dx_3 \tag{5}$$

Introducing the reduced stiffness tensor (for planar stresses) for the plate $\overline{R}_{ijk}$
and for the p-e wafer $\overline{R}^{p}_{ijkl}$, the expression (5) can be written in the following
form (see Ref. 3):

$$m_{ij} = -R^{F}_{ijkl}\varphi_{kl} - 2\int_{h}^{h+\alpha} x_3^2 \left\{ \left[\overline{R}^{p}_{ijkl} - h_{i,j}h_{kl}/c\right]\varphi_{kl} + h_{ij}V_{,3}/c \right\}dx_3 \tag{6}$$

where:

$$R^{F}_{ijkl} \quad = \quad \int_{-h}^{h}\overline{R}_{ijkl}\,dx_3$$

$$\varphi_{kl} \quad = \quad u_{3,kl}$$

h_{ij}—denotes the piezoelectric coefficients of the p-e wafer
c— denotes the dielectric constant of the p-e wafer
V—denotes the electric potential applied to the twin wafers
Since the thickness of the p-e wafers are very small comparing to the plate
thickness ($a \ll h$), the term x_3 can be approximated by $h + a/2$ in the last
component of the moment value in equation (6). Also considering $\varpi_p = \varpi$ what
give us stronger actuation effect the constitutive equation can be expressed as
follows:

$$m_{ij} = -R_{ijkl}\varphi_{kl} - 2(h + a/2)h_{ij}\Delta V/c \tag{7}$$

where $R_{ijkl} = R^{F}_{ijkl} + R^{P}_{ijkl}$, $R^{P}_{ijkl} = 2\int_{h}^{h+a} x_3^2 \left[\overline{R}^{P}_{ijkl} + h_{ij}h_{kl}/c\right]dx_3$,and ΔV
denotes the difference of potentials between the two p-e wafers (the upper and
the bottom wafer, respectively).

Let us now discuss the possibility of transformation of the equation (7) to
the following form (cf. Ref.5):

$$m_{ij} = -R_{ijkl}(\varphi_{kl} - \varphi_{kl}^{0}) \tag{8}$$

where   active distortions $\varphi_{kl}^{0} = \varphi_{kl}^{0}(\Delta V)$ describing the influence of the p-e
wafers on the distribution of moments will play the key role in the method of
determining the optimal position for actuators.

In the case of isotropic plate the non-zero components of the stiffness tensor
R^{F}_{ijkl} are the following (cf. Ref.12):

$$
\begin{aligned}
R^F_{1111} &= & R^F_{2222} &= & 2Eh^3/3(1-\nu^2) \\
R^F_{1122} &= & R^F_{2211} &= & 2E\nu h^3/3(1-\nu^2) \\
R_{1212} &= & Eh^3/3(1+\nu) &&
\end{aligned}
\tag{9}
$$

where E and ν denote the Young's modulus and the Poisson's coefficient, respectively.

In the case of uni-oriented p-e wafers the electromechanical constant takes the following form (choosing the axis x_1 for instance)

$$
h_{ij} = H\delta_{1i}\delta_{1j}
\tag{10}
$$

and non-zero components of the stiffness tensor R^P_{ijkl} are the following:

$$
\begin{aligned}
R^P_{1111} &= \tfrac{2}{3}\left[\frac{E_p}{(1-v_p^2)} + \frac{H^2}{c}\right] a(3h^2 + 3ha + a^2) \\
R^P_{2222} &= \tfrac{2}{3}\frac{E_p}{(1-v_p^2)} a(3h^2 + 3ha + a^2) \\
R^P_{1122} &= \qquad R^P_{2211} \qquad\qquad = \tfrac{2}{3}\frac{E_p}{(1-v_p^2)} a(3h^2 + 3ha + a^2) \\
R^P_{1212} &= \tfrac{1}{3}\frac{E_p}{(1-v_p^2)} a(3h^2 + 3ha + a^2)
\end{aligned}
\tag{11}
$$

where E_p and ν_p denote the Young's modulus and the Poisson's coefficient of the p-e wafer, respectively.

Substituting Eqs.9-11 to Eq.7 and comparing the result with Eq.8 the following conditions determining the active distortions φ^0_{kl} can be determined

$$
\begin{aligned}
R_{1111}\varphi^0_{11} + R_{1122}\varphi^0_{22} &= -\frac{(2h+a)H}{c}\Delta V \\
R_{2211}\varphi^0_{11} + R_{2222}\varphi^0_{22} &= 0 \\
R_{1212}\,\varphi^0_{12} &= 0
\end{aligned}
\tag{12}
$$

Finally, the tensor of active distortions calculated from the Eqs.(12) takes the following form

$$
\varphi^0_{ij} = -\frac{H(2h+a)}{c}R\Delta V
\begin{bmatrix} 1 & 0 \\ 0 & \xi \end{bmatrix}
\tag{13}
$$

where: $R = R_{2222}/(R_{1111}R_{2222} - R_{1122}R_{2211})$ and $\xi = R_{2211}/R_{2222}$. Therefore, mostly one component φ_{11} of the tensor of deformations can be controlled by our uni-oriented twin p-e wafers. As we can see, the influence of the p-e wafers on the behavior of the plate can be formally interpreted as the influence of some hypothetical fields of imposed distortions, taken into account through the modified constitutive relation (8).

3 Optimal Location for Piezoelectric Actuators

Let us formulate the problem of the optimal location for piezoelectric wafers assuming that:

- only n first modes of free vibration is going to be controlled

- each of these n modes is going to be controlled independently and therefore actuation controlling one mode is not going to affect other chosen modes,

- to maximize the total actuation efect it is assumed that the whole surface of the plate is covered by piezoelectric wafers.

Restricting considerations to the uni-oriented wafers (the control of maximal principal deformations can then be most effective) let us concentrate on the case of simple supported vibrating beam. However, the applied method can be easily generalized on the two-dimensional case of plate vibration, where uni-oriented p-e wafers should be distributed along the principal curvatures of deformation. Also, analogous application to quasistatical problems of shape control can be discussed. The modified constitutive relation (8) can be specified in the following form:

$$m = -R(\varphi - \varphi^0) \tag{14}$$

where:

$$
\begin{aligned}
R &= R^F + R^p \\
R^F &= \tfrac{2}{3}Eh^3 \qquad\qquad R^P = \tfrac{2}{3}E_p a(h^2 + 3ha + a^2) \\
\varphi^0 &= -\tfrac{H(2h+a)}{cR}\Delta V
\end{aligned}
\tag{15}
$$

Let $\varphi_i(x)$, $i = 1,...n$ be an eigenmode (a component of structural deformation to be controlled) determined by the free vibrational problem, and $D(x, y)$ be an influence function determining deformation $\varphi^R(x) = D(x, y)\varphi^0(y)$ caused in the location x of the beam by the unit distortion $\varphi^0 = 1$ generated in the location y. The best location problem can be now formulated as the following optimization problem (cf.Ref.7. for more explanations)

$$
\begin{aligned}
\sum_{i=1}^n \int_0^l a_i\varphi_i(x) \int_0^l D(x,y)\varphi_i^0(y)dydx &= \max \\
\bigwedge_{x\in[0,l]} \bigwedge_{\substack{i,j \\ i\neq j}} \varphi_i^0(x)\varphi_j^0(x) &= 0 \\
\bigwedge_{\substack{i,j \\ i\neq j}} -\delta \le \int_0^l \varphi_i(x)\int_0^l D(x,y)\varphi_j^0(y)dydx &\le \delta
\end{aligned}
\tag{16}
$$

where $\bigwedge_y \varphi_i^0(y) \in \{-1, 0, 1\}$, a_i—scalar weighting coefficients, n-number of chosen eigenmodes to be controlled, δ—the tolerance of satisfying the orthogonality problem. In order to do that we assume that on each subinterval $[y_m, y_{m+1})$ of the subdivision mentioned the functions φ_i^0 are constant and equal ξ_i^m (where $\xi_i^m = 1, -1, or\ 0$). That is why we can present any integral $\int_0^l D(x,y)\varphi_i^0(y)dy = \sum_{m=0}^{k-1} \int_{y_m}^{y_{m+1}} D(x,y)\varphi_i^0(y)dy$ (which appears in the objective function as well as in the constraints of the problem(16)) as a sum $\sum_{m=0}^{k-1} \xi_i^m \int_{y_m}^{y_{m+1}} D(x,y)dy$. Thus

$$\int_0^l \varphi_i(x) \int_0^l D(x,y)\varphi_i^0(y)dydx = \sum_{m=0}^{k-1} \xi_i^m \int_0^l \int_{y_m}^{y_{m+1}} \varphi_i(x)D(x,y)dydx$$

Under our assumptions the value of any particular integral

$$c_i^m := \int_0^l \int_{y_m}^{y_{m+1}} \varphi_i(x)D(x,y)dydx$$

can be pre-computed numerically quite easily. Then, the objective function of our main problem (16) can be presented as a linear function of the decision variables ξ_i^m $(i = 1, ..., n; m = 0, ..., k-1)$ of the following form

$$\sum_{i=1}^n \sum_{m=0}^{k-1} a_i c_i^m \xi_i^m \ where\ \xi_i^m \in \{-1, 0, 1\}. \tag{17}$$

In the same way we can turn the constraints of our original problem. The first set of constraints $\bigwedge_{x \in [0,l]} \bigwedge_{i \neq j} \varphi_i^0(x)\varphi_j^0(x) = 0$ for $(1 \leq i, j \leq n)$ can be presented in the equivalent form

$$\xi_i^m \xi_j^m = 0\ for\ \ m = 0, ..., k-1,\ 1 \leq i, j \leq n\ (i \neq j) \tag{18}$$

Similarly , the last set of constraints $(16)^3$ can be converted into a set of linear constraints of the form

$$\bigwedge_{\substack{1 \leq i,j \leq n \\ i \neq j}} -\delta \leq \sum_{m=0}^{k-1} c_i^m \xi_j^m \leq \delta \tag{19}$$

As we can see, our problem (16) has been reduced to the integer programming problem (with linear objective function). This new form is still very difficult to solve, because a huge number of nonlinear constraints (18). It is much more convenient to convert the problem presented above to the form of (0-1) linear programming problem (with linear objective function and linear constrains only). Next, we will use branch-and-bound method (see Ref.11) to solve it.

To do that we will introduce new decision variables ξ_j^{m+} and ξ_j^{m-} such that $\xi_j^m = \xi_j^{m+} - \xi^{m-}$ and $\xi_j^{m+}, \xi_j^{m-} \in \{0, 1\}$. After that we can present the final discretized form of our continuous problem (16) i.e.

condition $(16)^3$. The control function $\varphi_i^0(y)$ can take the values -1,0, or 1 what corresponds to the following three possible cases:

- $\varphi_i^0(y) = -1$ what means that the twin $p-e$ wafers in the location y control the mode i and generate the maximal available (negative) difference of potentials ΔV,

- $\varphi_i^0(y) = 0$ means that the twin p-e wafers in the location y do not control the mode i

- $\varphi_i^0(y) = 1$ means that the twin p-e wafers in the location y control the mode i and generate the maximal possible (positive) difference of potential ΔV.

The set of conditions $(16)^3$ means that each state of deformation $\varphi_j^R = \int_0^l D(x,y)\varphi_j^0(y)dy$ caused by actuators is orthogonal (with accuracy defined by the tolerance δ) to all eigenmodes φ_i, except one $(i = j)$. It follows from this condition that the control process can be uncoupled and therefore each chosen eigenmode can be damped independently. The tolerance δ will help us to satisfy the orthogonality condition for the discretized problem.

Let us now discuss the mechanical meaning of the optimization problem (16). The criterion $(16)^1$ requires maximization of a substitutive actuation corresponding to each mode separately with due weighting coefficient a_i and with eigenvectors φ_i playing the role of the (shape) weight functions. Constraints $(16)^2$ mean that in each point x of the beam only one piezoelectric twin wafer (related to one, defined mode control) can be bounded. Therefore, the whole surface of the beam can be finally covered (cf.$(16)^1$,$(16)^2$) by piezoelectric actuators divided on sections where each section corresponds to one actuator controling one eigenmode.

4 Discretization of the Best Location Problem

Let us consider the best location problem (16) again. In practice the analytic forms of the functions $\varphi_i(x)$ $(i = 1, ..., n)$ are unknown. We have only the values of these functions determined on the same regular grid of points $\{y_m\}_{m=0}^k$. We will assume that $0 = y_0 < y_1 < ... < y_k = l$. Similarly we have only values of the function $D(x,y)$ determined on the set of points (y_j, y_m) $(j, m = 0, ..., k)$ (i.e. on the product of the grid defined above). The analytic form of the function D is usually unknown as well. On the basis of the information available, we would like to substitute the original problem (16) by an auxiliary discrete optimization problem. After that we will try to solve this simplified problem numerically. Then we will obtain an approximate solution of the best location

$$\max \quad \sum_{i=1}^{n}\sum_{m=0}^{k-1} a_i c_i^m \xi_i^{m+} - \sum_{i=1}^{n}\sum_{m=0}^{k-1} a_i c_i^m \xi_i^{m-} \tag{20}$$

subject to:

$$\bigwedge_{0 \leq m \leq k-1} \qquad \sum_{i=1}^{n} \xi_i^{m+} + \sum_{i=1}^{n} \xi_i^{m-} \quad \leq \; 1$$

$$\bigwedge_{\substack{1 \leq i,j \leq n \\ i \neq j}} \qquad -\delta \; \leq \; \sum_{m=0}^{k-1} c_i^m \xi_j^{m+} - \sum_{m=0}^{k-1} c_i^m \xi_j^{m-} \; \leq \; \delta$$

$$\xi_i^{m+}, \xi_i^{m-} \in \{0,1\}$$

Without loss of generality we can assume that all of the components of the objective function and constraints are integer (we can multiply them by a big integer scalar if it is necessary). Under this assumption our maximization problem is a classical 0-1 linear programming problem and it can be solved directly. The algorithm we have used to solve this problem generally follows Zielinski's implementation of the Hammer and Rudeanu method. At every stage of it we have a partial solution and the corresponding (current) problem derived from the original one. In the partial solution, some variables are assigned fixed values (0 or 1) and the others remain free. The partial solution corresponding to the original problem has all the variables free. A partial solution is completed if all variables are fixed. Given a partial solution, we try to complete it. If there is a completion, we change the suplementary constraint $\sum_{i=1}^{n}\sum_{m=0}^{k-1} a_i c_i^m \xi_i^{m+} - \sum_{i=1}^{n}\sum_{m=0}^{k-1} a_i c_i^m \xi_i^{m-} \geq b_0$ (where b_0 is equal either to the value of the objective function at a feasible point or to the lower bound of it) and bactrack. If there is no completion, we also backtrack. In both cases, we go back to the last branching point and examine the new partial solution with the complementary value of the branching variable. As a result, either we find a maximizing point and maximum value of the objective function or the problem has no solution (which takes place when, for example parameter δ in (19) is too small). In [6] we explain how to determine appriopriate value of this parameter in the beginning (when the problem is formulated).

Of course the accuracy of the solution obtained this way depends on the diameter of the subdivision of the interval $[0, l]$. The smaller the diameter the better approximation of the solution of the original problem (16) can be obtained.

5 Numerical Example

The presented method of identification of optimal location for piezoelectric actuators is tested here on the example of free vibration of simply supported beam. Constraining the control process to the first three modes and assuming these

198

modes in the following form: $\varphi_1(x) = \sin(x)$, $\varphi_2(x) = \sin(2x)$, $\varphi_3(x) = \sin(3x)$ where $x = \pi\xi/l$, $\xi \in \langle 0, l \rangle$ and l denotes the length of the beam (cf. Fig.1), the optimal distribution of p-e actuators containing each mode separately is determined. The influence function $D(x, y)$ is approximated in this example by the following piece-wise-linear function:

$$D(x,y) = \left\{ \begin{array}{ll} -\frac{x}{\pi} + \frac{y}{\pi} + 1 & for \quad x \geq y \\ \frac{x}{\pi} - \frac{y}{\pi} + 1 & for \quad x < y \end{array} \right. \tag{21}$$

and the beam is divided on ten uniformly distributed finite elements.

The optimal result is strongly dependent on the weight coefficients a_i. Naturally, assuming $a_i = 1$ for i=1,2,3 the first mode is dominating and almost the whole beam is covered by the p-e wafer controlling this mode. Therefore, let us chose the following distribution of coefficients: $a_1 = 1$, $a_2 = 3$, $a_3 = 4$. The corresponding optimal distribution of actuators is presented in Fig.1.

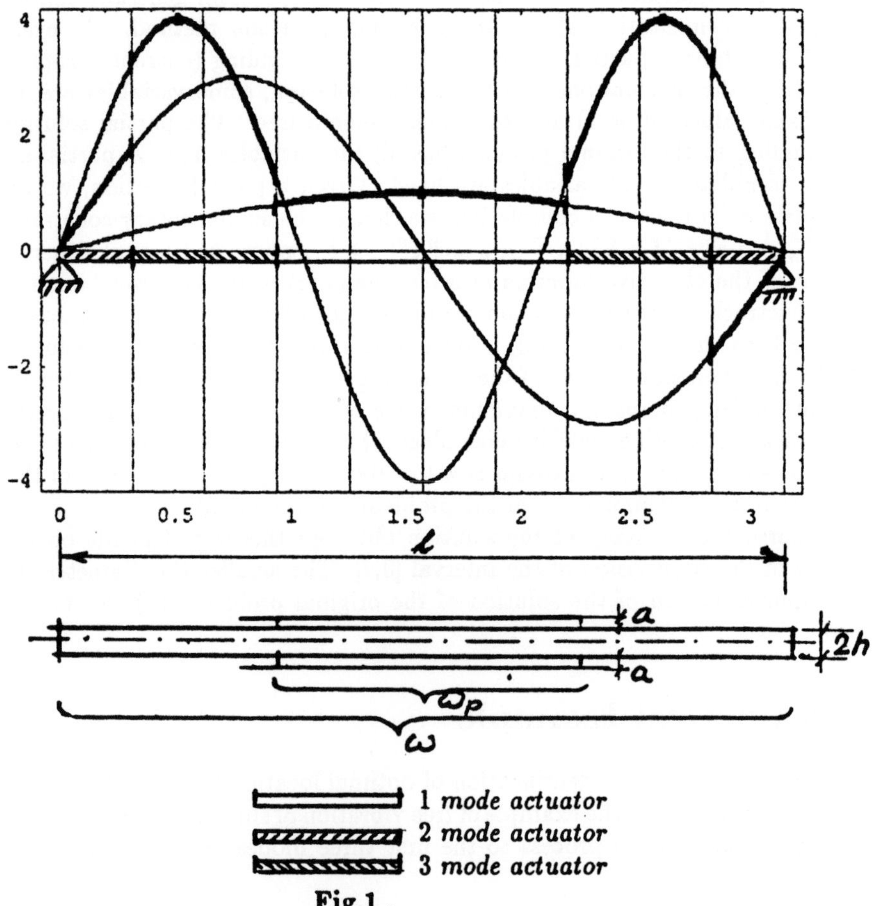

Fig.1 .

The scaled (due to the weight coefficients a_i) eigenmodes and the distribution of piezoelectric wafers containing each mode separately is shown. In real applications the proportion between coefficients a_i could be related to the corresponding eigenvalues (eigenfrequences).Further discussion of the problem presented above will be contained in the separate paper [7].

Acknowledgement. This work was supported by US Army Research Office grant DAAL03-92-G-0180 and Spanish Government (Direction General de Investigacion Cientifica y Técnica), Research Project No. PB 89-05-03.

References

[1] R. Burdisso and R. Haftka. *Optimal Locations of Actuators for Correcting Distortions in Large Truss Structures*, AIAA J., vol.27, NO.10, 1989, pp.1406-1411

[2] I.J. Chang and T.T. Soong. *Optimal Controller Placement in Model Control of Complex Systems*, J.Math.Anal.Appl. vol.75, 1980, pp.340-358

[3] Ph. Destuynder, I. Legrain, L. Castel, N.Richard. *Piezoelectric Wafers for Reducing the Structure Vibration*, Aerotechnique de Saint-Cyr Report, 1991

[4] R.T. Haftka. *Optimum Placement of Controls for Static Deformations of Space Structures*, AIAA J.,vol.22, 1984, pp.1293-1298

[5] J. Holnicki-Szulc, *Damping of Vibration by Actively controlled Initial Distortions*, J.Aerospace Eng.ASCE, vol4, No.1, 1991, pp.31-46

[6] J. Holnicki-Szulc, F. Lopez-Almansa, J.Rodellar. *Optimal location of Actuators for Active Damping of Vibration*, AIAA Journal (accepted, to be published in July 1993)

[7] J. Holnicki-Szulc, F. Lopez-Almansa, A. Mackiewicz. *Optimal distribution of Piezoelectric Actuators for Damping of Vibration*, (in preparation)

[8] A. Mackiewicz and D. Kozlowska. *Optimization Subroutines in Fortran* (in print)

[9] G.S. Nurre, R.S. Scofield and J.L. Sims. *Dynamic and Control of Large Space Structures*, J.Guidance Control Dyn., vol.7, 1984, pp.515-526

[10] A.E. Sepulveda and L.S. Schmit,jr. *Optimal Placement of Actuators and Sensors...*, Int.J.Num.Met.Eng., vol.32, 1992, pp.1165-1187

[11] S. Zielinski. *Solution of Linear Programming Problems in Zero one Variables*, Zastos. Mat.,vol.16,1977, pp.122-131

[12] O.C. Zienkiewicz, *The Finite Element Method* 3ed. McGraw-Hill 1977.

Optimal Discrete Design of Elastoplastic Structures

Algirdas Čižas and Stanislav Stupak

Department of Strength of Materials,
Vilnius Technical University,
Sauletekio al. 11, 2054, Vilnius, Lithuania

Abstract. A technique of an optimal design of elastoplastic framed structure is proposed if the axes of all the members of a frame are known and the optimal cross sections are to be chosen from a specific set. The optimality criterion is a minimum volume of a   structure. Various limit states of a structure are considered in the design problem: the ultimate values of displacements, strains and slenderness ratios. The   design problem is formulated as a problem of discrete linear programming and its solution is iterative.

Keywords. Discrete programming,   elastoplastic structure,   optimal design

1. Introduction

Beams and frames are usually assembled of commercially fabricated standard bars. A structure is seldom optimal if its cross sections are chosen according to the solution of any non-discrete design problem. The deviation from the optimum is more significant when in addition to the strength requirements the stiffness and stability requirements are included into a design problem. The strength of a structure depends on the section modulus, while its stiffness and stability are influenced by other geometric properties of the cross section (the moment of inertia and the radius of gyration). The relationships between these properties of each section are different. It is possible to take into account the requirements of the strength, stiffness and stability simultaneously only when all the cross sections of a structure are known. Therefore the best way of solving such a problem is an iterative search for a discrete solution.

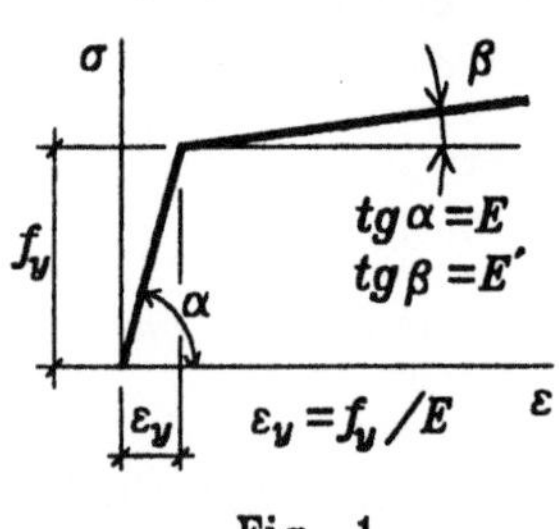

Fig. 1

Plane framed structures made of an elastoplastic material are under consideration. The stress-strain diagram of the behaviour of the material is bilinear (Fig. 1; the initial stage is elastic, the second perfectly plastic or linear strain hardening). All the parameters of the diagram are known.

The shape of the sections of the set is symmetric, all dimensions there are related to the section number or to its main design characteristic, namely the cross-sectional area. The plane of symmetry of any bar coincides with the bending plane in the structure.

A quasi static load is acting in the plane of the structure. The state of stress and strain of the structure should satisfy the restrictions of strains, displacements and slenderness.

2. Elastoplastic framed structure

2.1. Discrete analogue of the structure

The framed structure is divided into straight members connected with one another by nods. The load forces (and couple moments) are applied only to nods; any load distributed along the bar (beam) is reduced to several concentrated loads, and the bar is divided into several members so that the loads are applied to the nods between the members.

The design is based on the conventional assumptions: the displacements are small, the Bernoulli hypothesis is accepted, the influence of shearing stresses is negligible.

The stress state of any member k is defined by the axial force, which is constant along the whole member, and by two bending moments at the ends (i and j) of the member: $\tilde{Q}^{(k)} = (M_{ki}, M_{kj}, N_k)$ (a tilde above a symbol marks transposed vector or matrix).

The strain state of the member k is defined by the elastic (κ^e_{ki}, κ^e_{kj}) and plastic (κ^p_{ki}, κ^p_{kj}) parts of the curvature, by the elastic part of the axial longitudinal strain (ε^e_{0k}) and by its two plastic parts at the ends of the member (ε^p_{0ki}, ε^p_{0kj}): $\tilde{\tilde{q}}^{(k)} = (\kappa^e_{ki}, \kappa^e_{kj}, \varepsilon^e_{0k}, \kappa^p_{ki}, \kappa^p_{kj}, \varepsilon^p_{0ki}, \varepsilon^p_{0kj})$, or by the rotations of both ends (ϑ_{ki}, ϑ_{kj}) and the elongation of the central axis (δ_k): $\tilde{q}^{(k)} = (\vartheta_{ki}, \vartheta_{kj}, \delta_k)$. A relationship between those two vectors has been determined: $q^{(k)} = H^{(k)}\tilde{q}^{(k)}$, where the coefficient matrix $H^{(k)}$ may be calculated according to [1, 2].

The strain state of the whole structure is defined by the vector $\tilde{q} = (\tilde{q}^{(1)}, \tilde{q}^{(2)}, ..., \tilde{q}^{(k)}, ..., \tilde{q}^{(t)})$, where t is the number of members.

A non-linear relationship between the stress and strain states of the structure can be approximated by the linear law in a limited interval of the values of the internal forces:

$$q = GQ + \overline{G}, \tag{1}$$

where $\tilde{Q} = (\tilde{Q}^{(1)}, \tilde{Q}^{(2)},..., \tilde{Q}^{(k)},..., \tilde{Q}^{(t)})$, the coefficient matrix G and the vector $\overline{G}$ being calculated according to [2].

The displacements of the free nods of the structure are related to the strains by the equations of the geometric compatibility:

$$C\,u = q, \tag{2}$$

where $\tilde{u} = (u_1, u_2,..., u_{m})$ is the vector of the components of the nod displacements, m is the degree of freedom of the structure, C is the matrix of the geometric compatibility.

The state of stress of the structure is related to the load by the equilibrium equations:

$$\tilde{C}Q = F, \tag{3}$$

where $\tilde{F} = (F_1, F_2, ..., F_{m})$ is the vector of the load components.

If the mechanic and geometric characteristics of the members are known the matrices C, G, $\overline{G}$, H are known too. Then, if the load F is given, on solving the equations (1) - (3) the values of all the parameters of the stress and strain state can be calculated.

2.2. Discrete set of cross sections

Some members of the structure may be supposed to have the same cross sections, therefore $r \leq t$ different sections should be fixed in the solution of a design problem. A constraint for the selection of cross sections is fixed in the design problem: the cross section of any member of the group s ($s = 1, 2, ..., r$) may be chosen from the ordered set of sections $B^{(s)}$: $A_s \in B^{(s)}$; the sections are put into the set in the order of increasing area values: $A(B_i^{(s)}) < A(B_{i+1}^{(s)})$.

Any set $B^{(s)}$ is not numerous, because linear approximations of the relationships between the geometric characteristics of its cross sections (the moment of inertia I, the section modulus W, the radius of gyration i) should be acceptable within it:

$$I_s = a_s A_s + b_s, \qquad W_s = a_s' A_s + b_s', \qquad i_s = a_s'' A_s + b_s''. \tag{4}$$

If a large set of sections is taken for a general use, it should be decomposed into smaller sets with overlaps, for example, in such a manner:

$$\overbrace{B_1 \quad B_2 \quad B_3 \quad \underbrace{B_4 \quad B_5 \quad B_6 \quad B_7 \quad B_8}_{B^{(2)}} \quad B_9 \quad B_{10}}^{B^{(1)} \qquad\qquad\qquad B^{(3)}}$$

A solution is considered to be unreliable if it contains a boundary section (the first or the last one) from any set. Such a problem should be resolved after attaching the shady group to another ordered set of sections, the one in which the required cross-sectional area value is by the middle.

3. Optimal design problem

3.1. Mathematical analogue of the problem

The minimum volume of the structure $\sum\limits_{s=1}^{r} L_s A_s$ is the optimality criterion (there L_s is the total length of the members of the group s). The strength and stiffness of the structure and the stability of some (compressed) members are guaranteed by the restrictions for:

the longitudinal strains in the outermost fibers (or the plastic parts of strains)

$$\varepsilon_j \leq \varepsilon_j^0, \quad j = 1, 2, ..., 2\,t, \tag{5}$$

the slenderness ratios of the compressed members

$$\lambda_k \leq \lambda_k^0, \quad k = 1, 2, ..., t_1 \leq t, \tag{6}$$

the displacements of the structure nods

$$u_i \leq u_i^0, \quad i = 1, 2, ..., m_1 \leq m. \tag{7}$$

Other restrictions may also be included into the design problem.

The solution of an optimal design problem is iterative. For the first iteration the cross sections of all the member groups of the structure (with the characteristics A_s^*, I_s^*, W_s^*) are chosen from given sets on the basis of an engineer's experience or some approximate solution. When we have all the geometric characteristics we solve the equations (1) - (3) and estimate the values of the structure parameters to be restricted: the nod displacements u_i^* and their elastic parts u_i^{*e}, the strains of the outermost fibers ε_j^* and their elastic parts ε_j^{*e} and others. Using the relationships (4) and some assumptions we can express the constraints (5) - (7) in such a way that they will become the restrictions of the decision variables A_s. By relating the maximum strain at some cross section to the change of the section modulus of this cross section we rearrange the constraint (5):

204

$$-\frac{\varepsilon_j^{*e}}{W_s^*}\, a_s'\, A_s \;\leq\; \varepsilon_j^0 + \frac{\varepsilon_j^{*e}}{W_s^*}\, b_s' - \varepsilon_j^* - \varepsilon_j^{*e}; \tag{5a}$$

and by relating the slenderness ratio to the change of the radius of gyration we rearrange the constraint (6)

$$-\frac{\lambda_k^*}{i_s^*}\, a_s''\, A_s \;\leq\; \lambda_k^0 + \frac{\lambda_k^*}{i_s^*}\, b_s'' - 2\lambda_k^*. \tag{6a}$$

We suppose that: 1) the displacement of any nod is related to the moments of inertia of the cross sections of all the members; 2) this relationship is separable; 3) with the change of the moments of inertia the change of the nod displacement

is proportional to the change of its elastic part $\Delta u_i^e = \sum\limits_{s=1}^{r} \dfrac{\dot{\omega}_{is}}{I_s^*}\Delta I_s$, where $\dot{\omega}_{is}$ is

calculated as a sum (related to the group s) of the products of the elements from the row i of the matrix $(\tilde{C}KC)^{-1}\tilde{C}$ and of the corresponding elements of the

vector $Q^e = KC(\tilde{C}KC)^{-1}F$, because $u_i^{*e} = \sum\limits_{s=1}^{r}\dot{\omega}_{is}$ (K is the stiffness matrix

of the structure). We rearrange the constraint (7) in the following manner:

$$-\sum_{s=1}^{r}\frac{\dot{\omega}_{is}}{I_s^*}\, a_s\, A_s \;\leq\; u_i^0 + \sum_{s=1}^{r}\frac{\dot{\omega}_{is}}{I_s^*}\, b_s - u_i^* - u_i^{*e}\,, \tag{7a}$$

Thus, a mathematical analogue of an optimal design problem is formulated in the following way:

$$\left.\begin{array}{c} \sum\limits_{s=1}^{r} L_s A_s \;\Rightarrow\; \min, \\[2mm] (5a),\ (6a),\ (7a), \\ A_s \in B^{(s)},\quad s = 1,\, 2,\, \ldots,\, r. \end{array}\right\} \tag{8}$$

It is a linear discrete programming problem, a generalised expression of which is the following:

$$\left.\begin{array}{c} \sum\limits_{s=1}^{r} L_s A_s \;\Rightarrow\; \min, \\[3mm] \sum\limits_{s=1}^{r} c_{ks}\, A_s \;\leq\; d_k\,,\quad k = 1,\, 2\,,\, \ldots,\, p = 2t + t_1 + m_1\,, \\[3mm] A_s \in B^{(s)},\quad s = 1,\, 2,\ldots,\, r. \end{array}\right\} \tag{9}$$

3.2. Algorithm for solving the problem

Any attempt to check all feasible combinations of cross sections are not a reasonable way of solving the discrete programming problem (9). Thus we apply a special algorithm [3] based on the ideas of the Gomory method. The solution consists of two stages: a reduction of the searching region and a checking of the solutions in the reduced region.

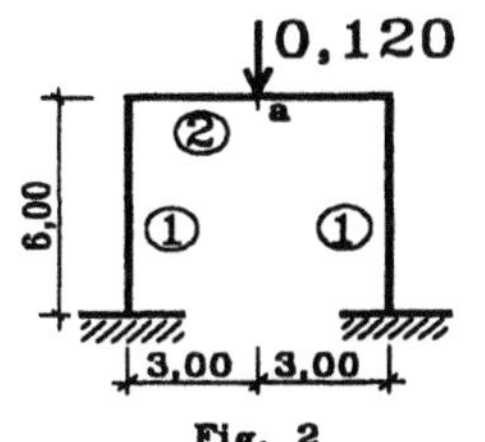

Fig. 2

An explanation of the algorithm will be illustrated by an example of a frame design (Fig. 2);

$$E = 200 \cdot 10^3, \quad E' = 5 \cdot 10^3, \quad f_y = 200,$$
$$\varepsilon^0 = 0,0026, \quad \lambda_1^0 = 120, \quad u_a^0 = 0,020.$$

A. *Reduction of searching region*

A1. A combination of the cross sections $A^{[0]}$ is chosen; there in the example $A^{[0]} = (A_1^{[0]}, A_2^{[0]}) = (12,0;\ 43,2)$. The relationships (4) acceptable in the proximity to $A^{[0]}$ are formulated.

A2. The step v with $A^{[v]}$ begins. We formulate the problem (9). By solving the problem without the requirement $A_s \in B^{(s)}$ we achieve the solution $A^{(v)}$. If the solution $A^{(v)}$ is close to the initial $A^{[v]}$ in the step $v = h$, it is acceptable and we pass to the item A4. There in the example (Fig. 3) $h = 3$, $A^{[3]} = (17,4;\ 72,6), \ A^{(3)} = (14,6;\ 68,8)$.

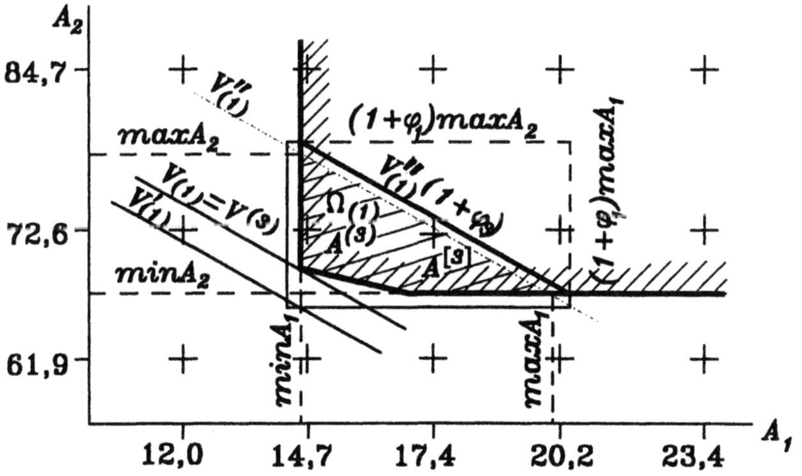

Fig. 3

A3. If any element of the solution $A^{(v)}$ is remote from the corresponding element of the initial $A^{[v]}$, the latter should be corrected to a new $A^{[v+1]}$, and we begin the step $v + 1$ (from A2).

A4. The value of the objective function $V_{(1)} = V^{(h)}$ is fixed. There in the example $V_{(1)} = V^{(3)} = 1200 \cdot 14{,}6 + 600 \cdot 68{,}8 = 58800$. We determine the factors $\varphi_1, \varphi_2, \varphi_3$ that depend on the deviations $(1 - A_s^{(h)} / A_s^{[h]})$ and the density of area values in the sets $B^{(s)}$.

A5. The iteration z of the final step begins. We reduce the searching region by two restrictions of the objective function value:

$$(1 - \varphi_1)V_{(z)} = V'_{(z)} \leq V = \sum_{s=1}^{r} L_s A_s \leq V''_{(z)} = (1 + \varphi_2)V_{(z)}. \tag{10}$$

In this way the whole region Ω of the sets $B^{(s)}$ is reduced up to the region $\Omega_{(z)}$.

There in the example: $\varphi_1 = 0{,}05$, $\varphi_2 = 0{,}10$ and $V'_{(1)} = 55860$, $V''_{(1)} = 64680 \; (z = 1)$.

A6. We solve a series of linear programming problems with the same constraints as in (9) - (10) but the objective functions are different:

$$A_s \Rightarrow \min, \quad A_s \Rightarrow \max, \quad s = 1, 2, \ldots, r.$$

After solving each pair of the problems we obtain the set $K_s^{(z)}$ that contains all the cross sections satisfying the inequalities:

$$(1 - \varphi_1)A_{s,\,\min} \leq A(B^{(s)}) \leq (1 + \varphi_1)A_{s,\,\max}. \tag{11}$$

If any set $K_s^{(z)}$ is empty we pass to the item A7. If all the sets $K_s^{(z)}$ are not empty the iterative procedure is completed ($z \equiv u$). There in the example when $z = 1$: $A_{1,\,\min} = 14{,}6$, $A_{1,\,\max} = 20{,}1$, $A_{2,\,\min} = 67{,}6$, $A_{2,\,\max} = 78{,}6$; and the sets of both cross sections are not empty, $n(K_1^{(1)}) = 3$, $n(K_2^{(1)}) = 1$, the iterative procedure is completed ($u = 1$).

We fix the region R for checking the discrete solutions. The number of solutions subject to checking is $n(R) = \prod_{s=1}^{r} n(K_s)$; there in the example $n(R) = 3$. We pass to the checking of the solutions, to the item B1.

A7. After determining a new value $V_{(z+1)} = V''_{(z)}$ a new iteration of the final step begins (from A5).

B. *Checking discrete solutions*

B1. The objective function value is calculated for each possible discrete solution:

$$\overline{V}^{(g)} = \sum_{s=1}^{r} L_s A_s^{(g)}, \quad g = 1, 2, \ldots, n(R).$$

All the values which satisfy the inequalities

$$V'_{(u)} \leq \overline{V}^{(g)} \leq (1 + \varphi_3) V''_{(u)} \tag{12}$$

are put into such an order that $\overline{V}^{(g)} \leq \overline{V}^{(g+1)}$. There in the example $\varphi_3 = 0{,}025$, the inequalities (12) are satisfied by the solutions $A^{(1)} = (14{,}7;\ 72{,}6)$ and $A^{(2)} = (17{,}4;\ 72{,}6)$; but they are not satisfied by the solution $A^{(3)} = (20{,}2;\ 72{,}6)$.

B2. A solution with a minimum value of the objective function is checked first of all. For this purpose a linear programming problem is to be solved:

$$\left. \begin{array}{c} x^{(g)} \implies \max, \\[2ex] -d_k\, x^{(g)} \leq \sum_{s=1}^{r} c_{ks}\, A_s^{(g)}, \quad k = 1, 2, \ldots, v. \end{array} \right\} \tag{13}$$

The solution $A^{(g)}$ is the optimal solution of the discrete programming problem (9) if $x_{\max}^{(g)} \geq 1$. There in the example $x_{\max}^{(1)} < 1$, $x_{\max}^{(2)} > 1$; thus, the optimal solution is $A^{(2)} = (17{,}4;\ 72{,}6)$.

B3. The solution $A^{(g)}$ is not within the more exactly defined feasible region of the problem (9) if $x_{\max}^{(g)} < 1$. If $g < n(R)$, we check the solution $A^{(g+1)}$ (from the item B2). If both $x_{\max}^{(g)} < 1$ and $g = n(R)$, there is no discrete solution in the feasible region $\Omega_{(u)}$ of the problem (9), and we should determine a new region $\Omega_{(z)}$ $(z = u+1)$ beginning from the item A7.

4. Conclusion

The applied algorithm of discrete programming ensures an optimal discrete design of an elastoplastic frame. The algorithm is suitable for other discrete programming problems with an objective function including only the variables that are to satisfy the condition $x_j \in B_j$.

208

References

1. **Čižas A.**: An algorithm for design of the elasto-plastic frame with due regard for the form of cross-sections (in Russian). Litovskii Mekhanicheskii Sbornik, 16, 104-113 (1976).

2. **Stupak S.**: Approximation of the law of the elastoplastic bending when sections of the frame are of any real shape (in Russian). Litovskii Mekhanicheskii Sbornik, 24, 39-43 (1981).

3. **Čižas A.**: An algorithm of discrete programming for optimization of framed structures compiled of the certain member set (in Russian). VIII Internationaler Kongress über Anwendungen der Mathematik in den Ingenieurwissenschaften. Berichte, 1. Weimar, 1978, pp. 171-176

Multicriterion Discrete Optimization of Space Trusses with Serviceability Constraints

Stefan Jendo[1] and Witold M. Paczkowski[2]

[1] Institute of Fundamental Technological Research PAS, Warsaw, Poland
[2] Civil Engineering Dept., Technical University of Szczecin, Szczecin, Poland

Abstract. The paper deals with discussion of discrete optimization problem in structural space-truss design. The cross-sectional areas of truss bars are taken as design variables. A catalogue of circular hollow sections for truss bars and a numbers of the series of types of cross-sections of bars are taken as discrete decision variables. The stress, local stability, displacement (design) constraints as well as technological and computational constraints are taken into account. The mass of truss bars including mass of joints as well as exploitation and maintenance costs are chosen as optimization criteria. A numbers of the series of types of cross-sections of bars t is also taken as a criterion of optimization. The sets of non-dominated (efficient) and compromise (Pareto optimal) solutions and the preferable solution for space truss are found. The results are presented in the form of diagrams.

Keywords. discrete optimization, multicriterion optimization, space truss design

1. Introduction

The double-layer space trusses are used for covering the exhibition, sport, shopping center and industrial halls [9]. Design in elastic range of such metallic structures is based on choosing the appropriate cross-sectional areas of bars to satisfy stress and local stability constraints. Next, the displacements of nodes under combined characteristic loads are calculated and compared with the allowable displacements occurring in codes [12]. If they are violated it is necessary to modify the structure by increasing the stiffness of structure.

The optimal structure should satisfy the limit state capacity and serviceability conditions [2,3]. Many papers in this area of research are only concerned with limit state capacity of structure. If the serviceability conditions are violated the structure is eliminated from considerations.

In this paper, a new method is presented that allows to tread concurrently the limit state capacity and serviceability conditions. It means that in the first stage of design, the minimal cross-sectional areas of truss bars satisfying the limit state capacity conditions are found. In the second stage the serviceability conditions are checked. If they are not satisfied the structure is modified. This modification can be done in a few ways.

2. Truss Design with Serviceability Constraints

The paper deals with optimal design of double-layer steel space trusses which can be used for covering of large span halls. The design codes require that the displacements of truss nodes under combined characteristic loads can not be larger than some specified permissible values as follows [12]

$$\Delta_1 \leq L_1/250, \qquad \Delta_2 \leq L/250, \qquad \Delta_3 \leq H/150 \qquad (1)$$

Usually, for small truss depth (h) it is easy to find the appropriate cross-sectional areas of truss bars to satisfy the stress constraints but the displacement constraint $(1)_1$ is often violated under given loads (Fig. 1).

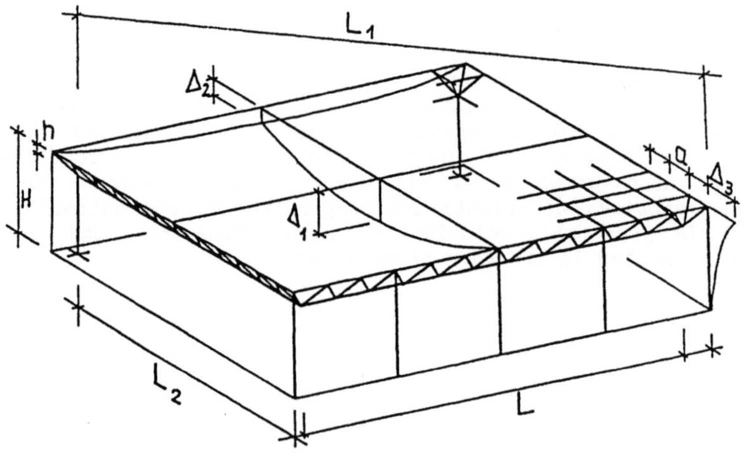

Fig. 1. Layout of double-layer space truss

The stiffness of double-layer space truss can be increased by the following means: i) increasing of depth of truss (h) [6,7], ii) increasing of distance between truss nodes (a) [6,7], iii) increasing a number of vertical supports [1], iv) changing supporting points from upper layer to lower layer [1], v) using a material with lower yield stress [7], vi) decreasing a numbers of the series of types of cross-sections for truss bars (t) [11], vii) decreasing a number of stiffness zones [10], viii) increasing the cross-sectional areas of truss bars with maximal stresses [7], ix) changing a truss topology (a type of bar net) [9], x) introducing an initial truss camber [8], xi) using the convex shapes of layers, xii) varying a shape of convex layers, xiii) applying truss joints with larger stiffness [4].

The most effective way to increase the truss stiffness D is to increase the truss depth (h) i.e.,

$$D = \frac{h^2 E A_u}{a(1 + A_u/A_l)}, \qquad (2)$$

211

where A_u and A_l are the cross-sectional areas of upper and lower
layer bars, E is the Young's modulus for steel. But, increasing the
truss depth (h) causes the increasing of hall height (H). Therefore,
the exploitation and maintenance costs are also increased. Some
structural systems, e.g., Mero system [9] have a fixed ratio between
truss bar lengths, i.e., it is not possible to increase the truss
depth (h) without increasing the distance between truss nodes (a) in
the upper and lower layers (Fig. 1).

In truss optimization problem each mean listed above concerns with
increasing a truss stiffness can be treated as independent decision
variable. The domain of feasible solutions can be increased taking
into account the stress, displacement and stability constraints.

3. Multicriterion Optimization of Space Truss

The paper deals with optimization of steel space truss used for
covering of exhibition halls (Fig. 2). The orthogonal double-layer
space truss is pinned supported at the corners of upper layer.
Circular hollow sections of truss bars are made from hot-rolled steel
with a strength of material f_d=210 MPa. The five single-layer
parabolic lattice-work domes are used as abatjours and they are
placed on the roof (Fig. 2).

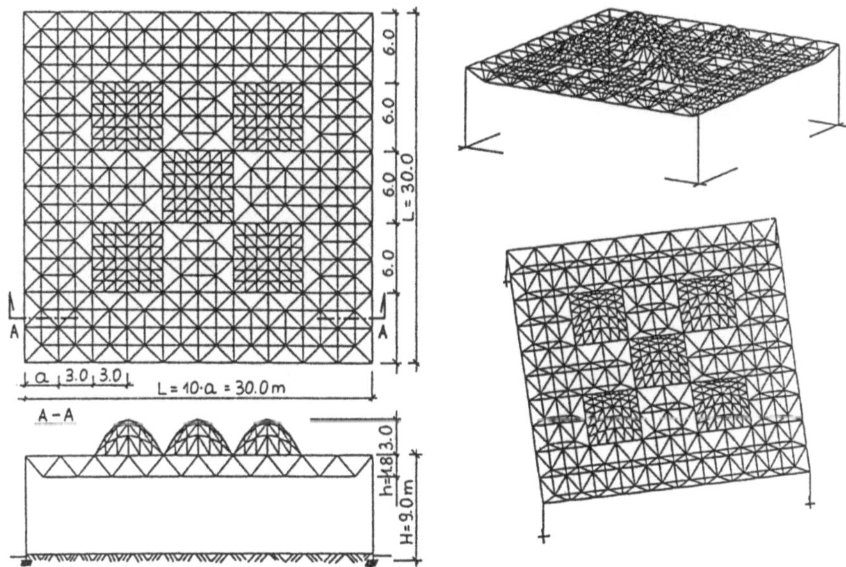

Fig. 2. Layout of covering of exhibition hall

The maximal loading condition which causes the maximal displacements
Δ_1 and Δ_2 (see Fig. 1) consists of the following loadings: the weight
of roof covering 0.5 kN/m^2, the snow loading 0.72 kN/m^2, the

technological loading 0.2 kN/m^2 and the concentrated loadings transmitted from abatjours to truss nodes shown in Figure 3. The supporting structure has a crucial influence on the horizontal displacement Δ_3. But Δ_3 is not constrained in this paper and it is not checked in static analysis of space truss. One quarter of space truss is only considered because of the symmetry of structure and loadings.

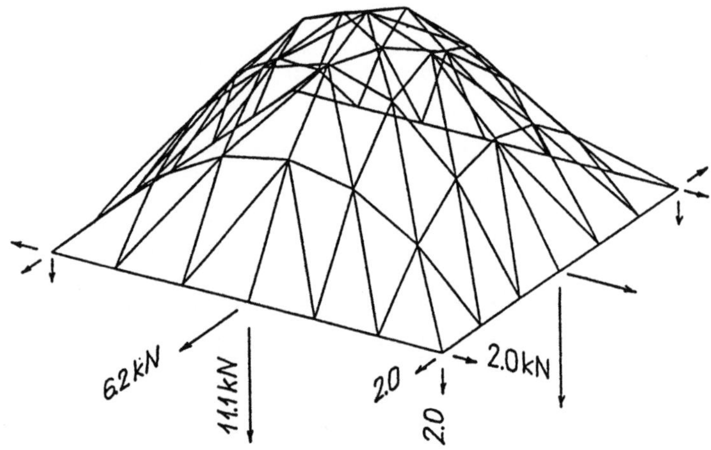

Fig. 3. The concentrated loadings transmitted from abatjours to double-layer truss nodes

The multicriterion optimization problem of space truss can be formulated as follows [5]. Find such catalogue of circular hollow sections for truss bars which minimizes the mass of truss for 1 m^2 of horizontal projection and a numbers of elements t in the catalogue i.e.,

$$f_1(x) = \frac{M_t}{(0.5L)^2} = \frac{4\gamma}{L^2} \, 1.1^{\frac{t}{t+0.1}} \sum_{k=1}^{k_b} A_k l_k , \qquad (3.1)$$

$$f_2(x) = t, \qquad (3.2)$$

where: M_t is the total mass of bars and joints, $\gamma=7850$ kg/m^3 is the bulk density of steel, A_k is cross-sectional area of k-th truss bar, l_k is the length of k-th truss bar, $k_b=168$ denotes the number of truss bars in one quarter of space truss, L=30 m is the truss span. The mass of truss joints for 1 m^2 of horizontal projection is taken into account by the empirical coefficient $1.1^{t/(t+0.1)}$ which is equal approximately 10% of mass of truss bars.

Vector of objective function has the following form

$$f(x) = [f_1(x),\ f_2(x)]. \tag{4}$$

Vector of decision variables has the form

$$x = [t,\ T] \in X, \tag{5}$$

where T is a catalogue of circular hollow sections for truss bars with a numbers of elements t (see Fig. 8).

The domain of feasible solutions is determined by the set of design, computational and technological constraints (Table 1).

Table 1. Optimization constraints for space truss design

Design constraints (6)	Technological constraints (7)
$\sigma_k^c = N_k\, m_{bk} / A_k \le f_d = 210$ MPa (6.1)	$A_k \in T \subset T_B \subset T_M$ (7.1)
$\sigma_k^t = N_k / A_k \le f_d = 210$ MPa (6.2)	$2.9 \le g \le 8.8$ mm (7.2)
$\lambda_k = l_k / i_k \le 250$ (6.3)	$31.8 \le D \le 244.5$ mm (7.3)
	$1 \le t \le 34$ (7.4)
$[K]_n \{\delta\}_n = \{P\}_n$ (6.4)	**Computational constraints (8)**
$\Delta_1 \le \sqrt{2}\, L/250 = 16.97$ cm (6.5)	$n \in \{1, 2, \dots, N\}$ (8.1)
	$1 \le N \le 10$ (8.2)
$\Delta_2 \le L/250 = 12.0$ cm (6.6)	$1 \le T \le 14$ (8.3)

The following notation in Table 1 is used: (σ_k^t) and (σ_k^c) are the tensile and compressive stresses in k-th truss bar, respectively, m_{bk} is the buckling coefficient for k-th compressed truss bar, $f_d = 210$ MPa is the strength of material, λ_k is a slenderness ratio for k-th compressed truss bar, i_k is a radius of inertia of cross-sectional area of k-th truss bar, $\{P\}_n$ is a loading vector, n is a number of iteration, T is the catalogue of circular hollow sections for truss bars, T_B is the basic catalogue of cross-sectional areas of truss bars and T_M is the metallurgical series of types of hot-rolled steel circular hollow sections, g and D are wall thickness and diameter of circular hollow section, respectively (Fig. 4).

The static analysis computer program is based on displacement method. The cross-sectional areas of truss bars are chosen automatically from a given the series of types of circular hollow sections. The local stability constraint (6.3) is taken into account. The internal forces in truss bars are determined. The appropriate cross-sectional areas of truss bars satisfying the constraints (6.1-6.4) and (7.1) are chosen in n-th iteration for the known stiffness matrix $[K]_n$ which was determined in the previous iteration.

The optimization process is terminated when the results of two following iterations are identical or maximal number of iterations N is reached. Next, the calculated loadings are replaced by characteristic loadings to find the actual displacements of truss nodes. If the displacement constraints (6.5) and (6.6) are not satisfied than a next catalogue is chosen and iteration process is repeated. It is assumed that the displacement constraints can be satisfied by choosing the appropriate catalogue of cross-sectional areas of truss bars.

Based on paper [11] the basic catalogue of cross-sectional areas of truss bars T_B is determinated. The basic catalogue T_B has the following feature: the ratio of cross-sectional areas A_{n+1}/A_n is constant and the corresponding critical N^c and limit N^t forces increase simultaneously. The maximal cross-sectional area is determinated on the base of maximal force which occurs in the truss with the same cross-sectional areas for each bar, i.e., t=1. The dimensions of cross-sections in the basic catalogue T_B and segment of metallurgical catalogue are shown in Figure 4.

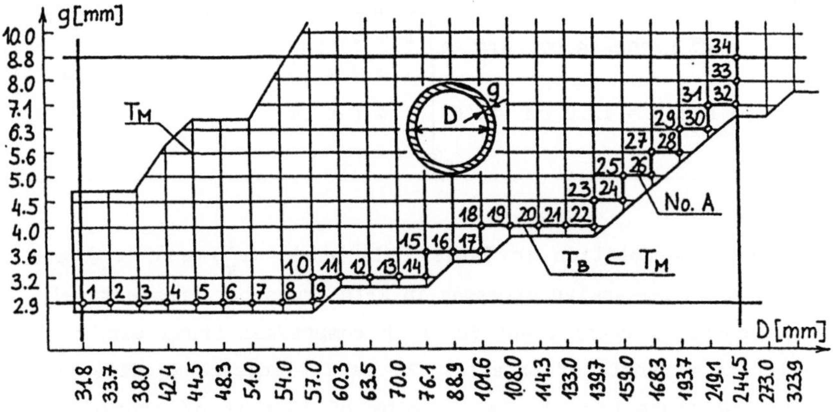

Fig. 4. The segment of metallurgical catalogue T_M and the basic catalogue T_B

Optimization process starts from a basic catalogue T_B. The displacement constraints for the assumed depth of double-layer space truss h=1.8 m are violated (Fig. 5). The sequential catalogues are created from the previous one by erasing every second cross-section (Fig. 6). The displacement constraints are only satisfied by the catalogues with a numbers t≤4 (Fig. 5). Among these catalogues the four non-dominated catalogues are chosen.

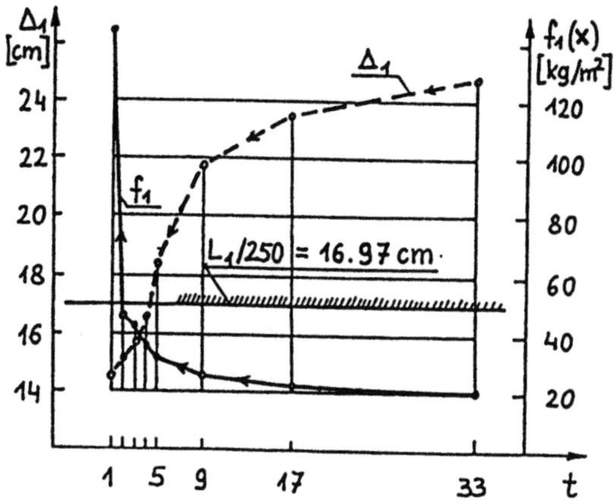

Fig. 5. Diagrams of the displacement Δ_1 and truss mass $f_1(x)$ versus a numbers of of elements t in the catalogue T

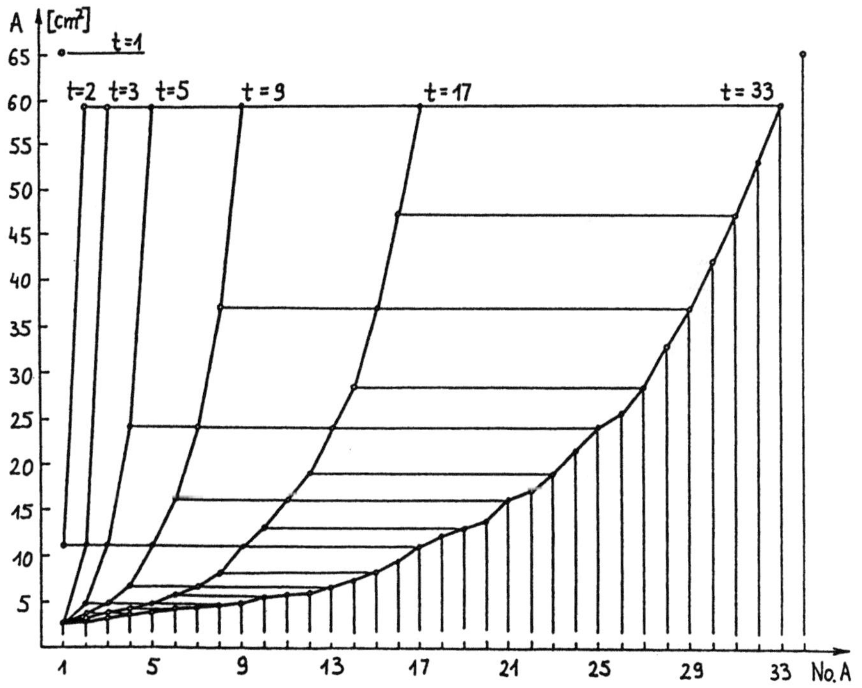

Fig. 6. Creation of catalogues by erasing every second cross-section (No. A is the number of cross-section in the catalogue T with a numbers t)

The space of decision variables $\mathcal{A}$ and the objective space $\mathcal{B}$ are shown in Figure 7.

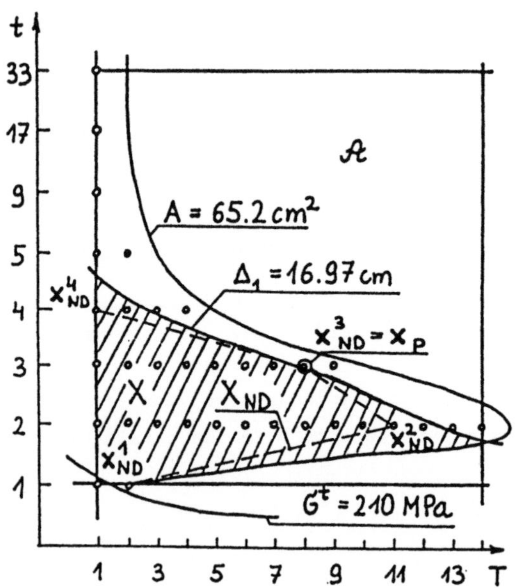

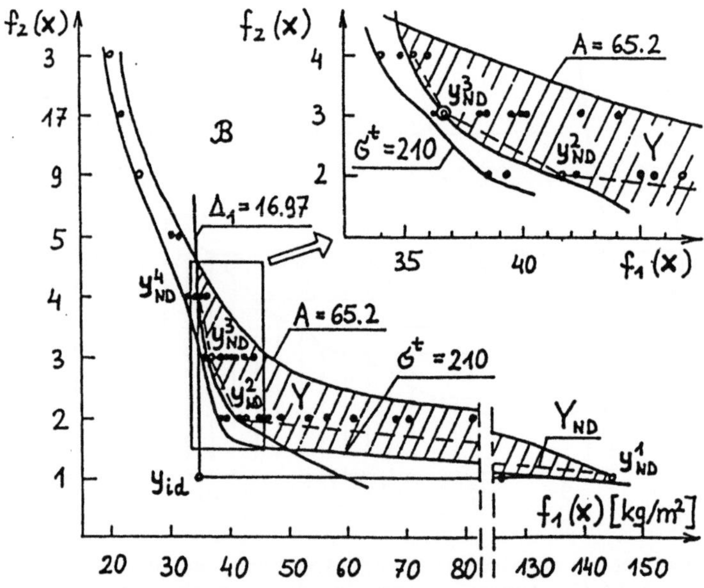

Fig. 7. The spaces of decision variables $\mathcal{A}$ and objectives $\mathcal{B}$

The active constraints (6.2), (6.5), (7.1), (7.4) and (8.3) determine the domain of feasible solutions $X \subset \mathcal{A}$. The image of the set of constraints listed above in objective space creates the attainable domain $Y \subset \mathcal{B}$. The discrete solutions $x \in X$ and the corresponding points $y \in Y$ are marked by circles. The non-dominated (efficient) points y^i_{ND}, $i=1,\ldots,4$ and the corresponding Pareto optimal (compromise) solutions $x^i_{ND}=[f(x)]^{-1}$ are linked by dash lines.

The curves $\Delta_1(x)$ and $f_1(x)$ for catalogues T with a numbers $t=2$ and $t=3$ are shown in Figure 8. The numbers corresponding to circular hollow sections occurring in basic catalogue T_B are given for each the series of types of cross-sectional areas of truss bars.

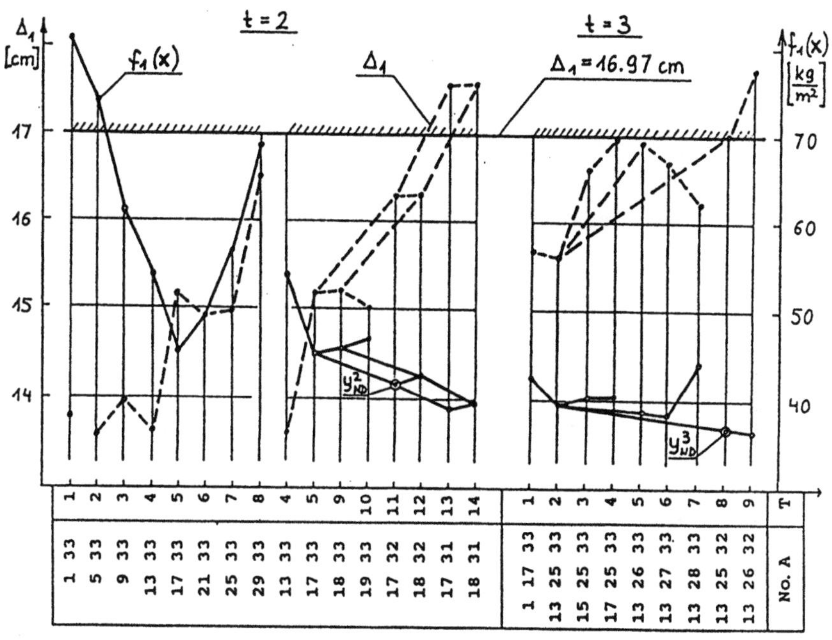

t = 2:

T	1	2	3	4	5	6	7	8	4	5	9	10	11	12	13	14
No. A	33	33	33	33	33	33	33	33	33	33	33	33	32	32	31	31
	1	5	9	13	17	21	25	29	13	17	18	19	17	18	17	18

t = 3:

T	1	2	3	4	5	6	7	8	9
No. A	33	33	33	33	33	33	33	32	32
	17	25	25	25	26	27	28	25	26
	1	13	15	17	13	13	13	13	13

Fig. 8. The curves $\Delta_1(x)$ and $f_1(x)$ for catalogues T with a numbers of elements $t=2$ and $t=3$

The controlled enumeration method based on structural design knowledge is used to find the set of compromise solutions. The performance of this method is schematically shown in Figure 9. The searching for minimum of $f(x)$ starts from the fixed $x_1 = x_1^0$. Next, after finding the optimal value for $x_2 = x_2^1$, the searching is continued along x_1 for the fixed $x_2 = x_2^1$ and so on. A more accurate solution can be found by reducing a mesh density of net.

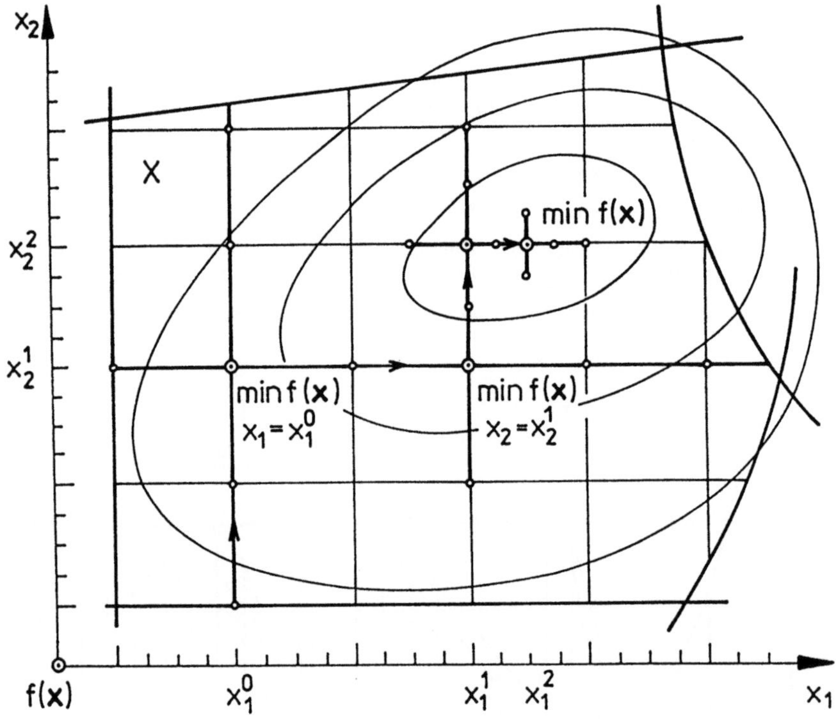

Fig. 9. The performance of the controlled enumeration method

One can choose a preferable solution x_p from the set of compromise solutions X_{ND} using a utility function [3], i.e.,

$$U(x_p) = \min_{x \in X_{ND}} U(M_t, x) = \min_{x_{ND} \in X^1} f_1(x)(1 - 0.4e^{-0.35t}), \qquad (9)$$

The coefficient $C(t)=(1 - 0.4e^{-0.35t})$ occurring in the utility function (9) and shown in Figure 10 describes the influence of labor cost concerning with a numbers of elements t in the catalogue of circular hollow sections T. .

The preferable truss is build from the catalogue of cross-sectional areas of truss bars with a numbers t=3. The mass of truss is $f_1(x)=36.64$ kg/m^2 and the displacement of central truss node $\Delta_1=17.02$ cm. The displacement of central node is 0.3% larger than the allowable one but it is still less than the permissible violation which is equal 2 %.

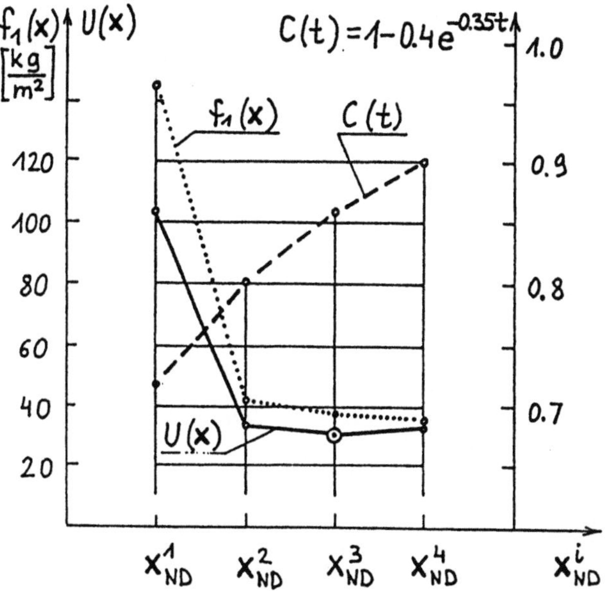

Fig. 10. The diagram of utility function and the preferable solution $x_p = x^3_{ND}$

4. Conclusions

The domain of feasible solutions should be determinated not only by stress constraints but also by displacement (serviceability) constraints. Optimum design of truss structure with respect to serviceability constraints requires the another algorithm than one concerns with stress constraints only. The displacement constraints can be satisfied by increasing the stiffness of the structure. The displacement of truss nodes can be decreased by a few independent means listed in the introduction. Optimum design of trusses with respect to serviceability constraints leads to increasing of truss mass but decreasing of labor cost and exploitation cost of object. Multicriterion optimization approach allows to find the compromise and non-dominated solutions. The preferable solution can be chosen taking into account the labor costs.

References

1. Bródka, J. (Ed). *Steel Space Structures,* (in Polish) Arkady, Warsaw 1985.
2. Grierson, D.E., and Chiu, T.C., "Synthesis of frameworks under multilevel performance constraints, *Solid Mechanics Division, University of Waterloo, Canada,* No. 172, 1982.
3. Grierson, D.E., and Lee, W.H., Optimal synthesis of frameworks under elastic and plastic performance constraints using discrete sections, *J. Struct. Mech.* 14, No.4, 401-420, 1986.

4. Grierson, D.E., and Xu, L., Design optimization of steel frameworks accounting for semi-rigid connections, *Lecture Notes: Vol.3, NATO/DFG ASI "Optimization of Large Structural Systems"*, 139-158, 1991.
5. Jendo, S. "Multiobjective optimization", in: *Structural Optimization, Vol.2: Mathematical Programming,* Save, M. and Prager, W. (Eds), Plenum Press, New York 1990, pp. 311-342.
6. Jendo, S. and Paczkowski, W.M. "Multicriteria Discrete Optimization of Large Scale Truss Systems", *Structural Optimization,* Vol.6, No.3, 1993 (in print).
7. Karczewski, J.A., Łubiński, M. and Paczkowski, W.M. "Optimization of some type of space trusses according to economic criteria" (in Polish), *Archives of Civil Engineering (Archiwum Inzynierii Ladowej)*, 31, 1/2, 1985, pp.57-80.5.
8. Karczewski, J.A., Niczyj, J and Paczkowski, W.M. "Multicriterion selection of spatial truss crown", *Proc. of the IASS-MSU Int. Symp. "Public Assembly Structures from Antiquity to the Present"*, Istambul 1993, pp.461-470.
9. Mengeringhausen, M. *Raumfachwerke aus Staben und Knoten,* Bauverlag GmbH, Wiesbaden und Berlin, 1975.
10. Paczkowski, W.M. "The multioptimal choice of stiffness zones in space truss supported at corners"(in Polish), in *Proc. of the IX Polish Conf.on Computer Methods in Mechanics,* Kraków-Rytro, Poland, 1989, pp.859-866.
11. Paczkowski, W.M. "The bars type range discrete polioptimization of space trusses" (in Polish) *Annales Scientarum Stetinenses,* Vol.6, 1, No.2, Ossolineum 1991, pp. 83-105.
12. PN-90/B-03200 *Polish Code for Steel Structures, Design and Rules* (in Polish), Code Publishers, ALFA, 1990.

Symbolic Computations Approach in Controlled Enumeration Methods Applied to Discrete Optimization

Mariusz PYRZ
Ecole Universitaire d'Ingénieurs de Lille, USTL-LML ,Département Mécanique,
F-59655 Villeneuve d'Ascq Cedex, France

Abstract: A coupling symbolic and numeric computation approach in the discrete optimization of problems with linear objective function is presented. An enumeration method according to the non decreasing values of the objective function is supplied with an additional symbolic module containing the domain oriented knowledge. This module eliminates the constraints verification for "non-promising" propositions and enables a significant reduction in number of design variables variants that must be checked for feasibility to find the global minimum. An optimization example for a simple truss structure and a few effective heuristic rules are presented.

Keywords. knowledge-based optimization, discrete optimization

1 Introduction

The main idea of the controlled enumeration methods applied to the discrete optimization consists of using such algorithms which give the optimal solution by partial enumeration, without checking all feasible variants. Generally an ordered sequence of design variable vectors is generated using some combinatorial techniques. For these vectors the constraints of the problem are checked until a criteria for solution is satisfied. The number of design variable sets that must be verified to find the global optimum can be very significant in real problems and implies a considerable time of calculus. A piece of knowledge can often substantially reduce the computational effort. This statement motivates the present paper.

The enumeration technique would be more efficient if "non promising" candidates were eliminated from the checking constraints procedure. However a knowledge about the problem to solve is necessary to remove "a priori" unrealistic, redundant and infeasible variants. The Artificial Intelligence techniques enable powerful processing of symbolic data. They can be used to represent the knowledge that will enhance the controlled enumeration algorithms.

222

The discrete optimization of problems with linear objective function is considered in the paper. The problem-oriented knowledge is used to eliminate a priori some candidates issued from the controlled search procedure. The enumeration module, the knowledge-based module and the numerically oriented module for constraints checking are separated. The truss optimization example demonstrates the potential of the proposed approach applied to engineering problems. A significant reduction in number of design variables variants that must be checked for feasibility to find the global minimum has been observed. The presented method can be included into the class of knowledge-based optimization algorithms, joining advantages of knowledge-based systems and "traditional" system of numerical analysis.

2 Formulation of the problem

The optimization problem considered here can be generally formulated for N discrete design variables as follows

$$(1) \quad F = \sum_{i=1}^{N} c_i x_i = cx \longrightarrow \min \quad ,$$

$$(2) \quad x \in S \ ,$$

$$(3) \quad x_i \in \{ \ x_i^1 \ ; x_i^2 \ ; x_i^3 \ ; \ ... \ ; x_i^{mi} \ \}, \quad i = 1,...,N \quad ,$$

where F is the linear objective function to be minimised, x is the vector of N discrete design variables, c is the vector of N constant real coefficients characterising components of the objective function, S is a set of feasible solutions determined by the arbitrary constraints. The design variables x_i (i=1,...,N) are to be chosen from a finite set of m_i feasible discrete values (3), where m_i stands for the number of discrete values which can be taken by x_i-th variable.

The solution of the problem under consideration proposed in the present paper joins a combinatorial algorithm of controlled enumeration, a knowledge based module and a constraint verification part.

3 Enumeration method according to the non-decreasing values of the objective function.

A version of the enumeration method according to the non decreasing values of the objective function (Iwanow 1990) has been adapted to the discrete optimization problem (1-3).

The idea of the method is to find an ordered sequence of design vectors

$$(4) \quad x_1; \ x_2; \ x_3; \ ...$$

according to the non-decreasing values of the corresponding objective functions

$$(5) \quad F_1 \leq F_2 \leq F_3 \leq ...; \quad F_i = F(x_i).$$

If min is the smallest natural $n \in N$, for which the condition $x \in S$ is satisfied, then the solution of the minimization problem is

$$(6) \quad x_{opt} = x_{min}; \quad F_{opt} = F(x_{opt}).$$

For ordered coefficients c_i of the objective function the algorithm constructs recursively a tree structure, assigning a unique value of F and x to each vertex. The suitable vertices (or all groups of vertices) are examined making use of the monocity properties of the created graph. At each step all variants equivalent with respect to the objective function are found. The search can be started from any arbitrary value, stated as a lower bound for objective function values to be generated in non-decreasing order. The algorithm does not require much memory and is more efficient from the numerical point of view than another algorithm of enumeration based on the same principle used by the author for similar optimization problems (Pyrz 1990).

4 Coupling symbolic and numeric computations

In the enumeration method described in the previous chapter the constraints of the problem have to be checked for all subsequently generated design variables vectors x_1 x_2 x_3 ... until reaching a x_{min} satisfying (2). The optimization procedure would be more efficient if one eliminated non promising candidates without checking them numerically for feasibility. A lot of design variable sets can be removed "a priori" thanks to an understanding of the expected results and a knowledge about the problem to solve. Many decisions of skipping non-promising variants can issue from technology, manufacturing, mechanical or other problem oriented properties that can be stated without calculus. The different forms of incorporated "domain oriented" knowledge can improve the efficiency of the optimization process.

The main idea of the proposed approach is to create three separate modules corresponding to different levels of processing. The explicit separation of modules is a natural consequence of the application of the controlled enumeration method.

4.1 Enumeration module

The enumeration method algorithm presented in the previous chapter is the first part of the program. It needs only the coefficients c and discreteness conditions of type (3) defining sets of available discrete values. As a result a sequence of design variable vectors corresponding to the non-decreasing values of objective function is generated. For these variants the constraints of the problem have to be verified.

4.2 Knowledge module

A separate symbolic processing module acting as a filter between generation of candidates and verification of the constraints is the second part of the program. This module removes the candidates from the checking constraints procedure if they are considered infeasible on the basis of the "problem oriented" information contained in the knowledge base. The knowledge base consists of domain facts, rules and heuristics associated with the problem and accompanied by a reasoning technique. An *IF condition THEN action* rule-based production system representation has been chosen. It accesses many facilities in manipulating knowledge, provided by symbolic computing. Its content is represented in a pseudo-natural language of the problem. It is easy to change or actualize thanks to the declarative characteristics: one should only precise what the system has to know. An adequate strategy of filtering has to be defined so as none of "good" candidates (with respect to some corresponding criteria) be skipped.

4.3 Constraints module

The check for feasibility part is generally connected with heavy numeric processing. In this module the design variable variants issued from the knowledge module are subsequently checked for feasibility. The first design vector satisfying the constraints (2) of the problem gives the optimal solution.

5 Problem oriented knowledge

The symbolic processing module is created to limit a combinatorial explosion of possible discrete design variables variants that must be verified to find the optimal solution. It disposes a very specific knowledge connected with the characteristics of the problem to solve, however some general rules or heuristics can be included.

The domain oriented knowledge useful in the optimization of engineering problems can issue from:

- technological, mechanical behaviour or manufacturing constraints (e.g. some combinations of design variables are not allowed because of manufacturing limitations or given target technology; there exists some preferable variants for a given type or range of design variables etc.);
- rules (or heuristic formalism) of modelling the considered class of structures;
- database of previous results calculated for similar problems;
- the priorities between equivalent (with respect to the objective function value) solutions or a criteria to choose the best possibility among the equivalent ones (some sets of design variables can be preferable in comparison with the others however all of them have the same value of the objective function);
- the design variables interrelations difficult to include into numerically represented constraints (utilities aspect, functional specifications, simplicity, aesthetic functions, visual characteristics, etc.).

6. Illustrative example

6.1 Formulation of the truss optimization problem

The weight minimization of a 10 bar truss structure (Fig.1) is presented to illustrate the proposed approach. The design variables are cross-sectional areas of 10 bars. The inequality constraint has been imposed on the vertical displacements of nodes as an upper limit of 2 in. The maximal allowable stress has been limited to $2.5*10^5$ psi. For the dimensions and material characteristics presented on Fig.1 the structure has been investigated for different applications of two vertical forces $P=10^5$ lb (including the case of loading a node by 2P) acting in downward direction. The number of design variables variants that must be checked for feasibility to find the global optimum has been studied. The results from the "standard" enumeration method application and from the knowledge-based enumeration approach have been compared.

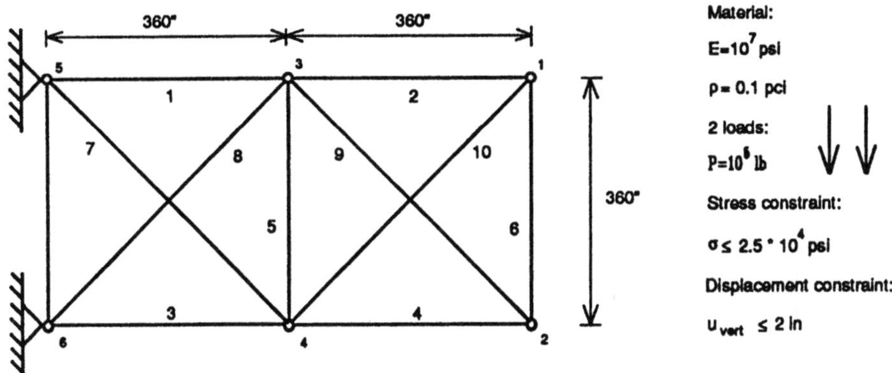

Fig.1 Ten-bar truss

6.2 Heuristic rules

The following heuristic rules have been formulated on the basis of the static of the structure and the constraints imposed on displacements . They can be viewed as a predimensionning of the cross sectional areas of the bars, where two sections A_5 and A_6 have been chosen as reference values. The rules take into account the loads positions and tend to prevent the displacement limit excess. For loading cases classified as leading to relatively small displacements, a combination of areas of bars 1,3,7,8 ("the left part of the structure") has to be increased with respect to A_5. If displacements are classified as big, "the right part of the structure" has to be additionally reinforced i.e. sections A_2 and A_{10} or A_4 and A_9 have to be greater than the section of reference A_6.

IF   *loads_in_nodes(1,1)* or *loads_in_nodes(2,2)* or *loads_in_nodes(1,2)*
THEN   *displacements_are_big.*

IF   not(*displacements_are_big*)
THEN   *displacements_are_small.*

IF   *displacements_are_small*
THEN   *reinforce_left_part.*

IF   *displacements_are_big*
THEN   *reinforce_left_part* and *reinforce_right_part.*

IF   *reinforce_left_part*
THEN $(A_1>A_5$ and $A_3>A_5$ and $A_7>A_5)$ or $(A_1>A_5$ and $A_3>A_5$ and $A_8>A_5)$

IF   *reinforce_right_part*
THEN $(A_2>A_6$ and $A_{10}>A_6)$ or $(A_4>A_6$ and $A_9>A_6)$.

6.3 Results for 4th-element catalogues

The above rules have been included to the symbolic module of the algorithm and the optimal solutions have been found for different loading cases. An additional condition on $A_5 <15.0$ and $A_6 <15.0$ has been included on reference values. The 4-th-element catalogue of discrete cross-sections {12.0 19.0 27.0 36.0} (in square inches) has been chosen for all design variables. The full survey of all possible variants would need the checking of $4^{10}=1048376$ propositions. The results compared with those obtained without the knowledge-based module are presented in Table 1. The loading case is described by two numbers of nodes, where two forces $P =10^5$ lb are acting. The number of equivalent solutions is given but only the number of the first optimal solution is referenced.

Loaded nodes	2 4	2 2	1 2
Optimal weight [lb]	6769.1435	9747.5232	9507.8764
Nr of equivalent solutions	2	1	4
Nr of checked variants for "standard" enumeration	4347	490236	410558
Nr of checked variants for knowledge-based enum.	150	15974	13315

Table 1. Comparison between the number of variants that must be checked
to find the optimum for different loading.

6.4 Conclusions

It shows from the presented examples that the knowledge-based approach implies an enormous reduction in the number of variants that must be checked even for a very simple knowledge base. The average economy with respect to the "standard" enumeration version was about 96,5%. It means that the constraints for only about 3,5% of generated variants needed to be verified to reach the optimum.

In the prototype version of the discrete optimization method proposed in the paper the enumeration and constraint checking modules have been coded in Fortran 77 while the knowledge-based part has been written in Turbo Prolog (version 2.0 of Borland).

7. Final remarks

A prototype knowledge-based discrete optimization method has been presented. A domain-specific knowledge is an active component of the algorithm and has been used to eliminate a priori "incorrect" design variables sets. The potential of symbolic computations applied to problems of discrete optilmization has been emphasised, however an additional development work is necessary. The linearization methods existing for the non-linear optimization (e.g. Walukiewicz 1991) enables the above general formulation to be applied to a much larger class of discrete problems. It is hoped that the knowledge-based approach used in conjunction with numerical techniques would considerably enhance the efficiency of optimization procedures. The further studies of the knowledge acquisition, formalization of heuristics or automated learning from the database of optimization examples are indispensable to code and utilise this knowledge properly.

References

Iwanow,Z.1990:An algorithm for finding an ordered sequence of a discrete linear function. Control and Cybernetics.19,3-4,129-154

Pyrz,M.1990:Discrete optimization of geometrically nonlinear truss structures under stability constraints. Structural Optimization. 2,125-131

Walukiewicz,S.1991:Integer programming, PWN-Polish Scientific Publishers, Kluwer Academic Publishers,Warszawa

General P-Δ Method in Discrete Optimization of Frames

Andrew T. Janczura [*]

ABSTRACT

The general P-Δ approach (GΔPN) for the finite element method (FEM) in discrete optimization of frames has been presented in this paper.

It has been shown that P-Δ effects and suitable solutions algorithms comes from some modifications of the standard secant stiffness method for the second order theory (MNII) in general.

The GΔPN method is accurate approximation of the MNII with opposite to a standard (SΔP) and a modified (MΔP) P-Δ methods. Moreover SΔP and MΔP methods are use to be applied to the rectangular frame. with oposite to the GΔPN.

The GΔPN approach is highly recomended for discrete optimization models of frames because is much effective as fastest in computer time, enough accurate and convergent even in critical area then the MNIIN.

With respect to structures, especially rectangular frames ,the GΔPN method, its first linearization (GΔPL) , standard and modified P-Δ methods , full nonlinear MNII and its linearization NMIIL has been compared .

One representative example of the frame has been chosen. A behaviour of GΔP methods in noncritical and critical area of loading with respect to their convergence has been considered.

A numerical example shows that all P-delta methods with comparison to the exact second order analysis gives satisfied results in noncritical area but received by the GΔP methods are the best. Moreover from the P-Δ approaches, only the general P-Δ is convergent in a critical area of loading.

1. Discrete optimization model

The optimization model of a steal frames is defined as nonlinear, discrete programming with constraints.

The design variables are defined by the linear dimensions of the cross sections and they have a discrete character.

One type of the cross sections is shown on a figure 1.

Suitable identification of the variable dimensions for the cross sections of all members, creates variable vector $\underline{X}$.

Phd.Civ.eng., Msc. math., Computer Methods In Civil Engineering, Building Engineering Institute, Technical University of Wroclaw, tel:(071)20-38-54,(071)20-38-82, Email:ATJAN@PLWRTU11.BITNET

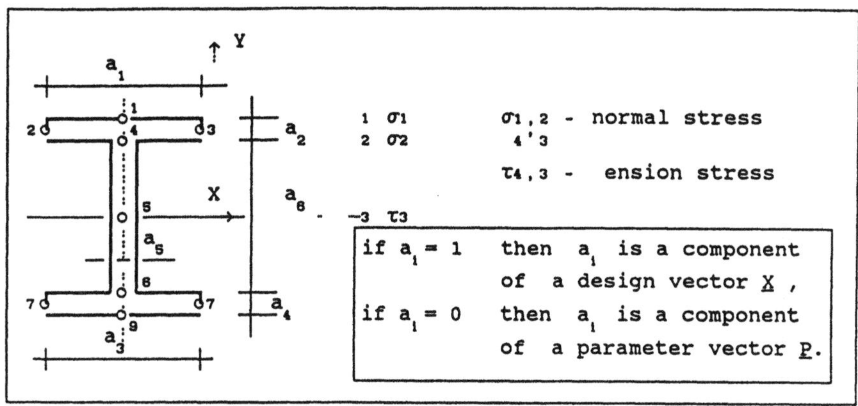

Fig. 1. T-cross section with the design variable, parameter
definition and calculations points.

This identification is equivalent to existence of one permutation
matrix B which leads to disclose the $\underline{X}$ and $\underline{P}$ vectors [8]:

$$B \cdot \underline{X} = \{ \underline{X} , \underline{P} \} \tag{1}$$

where $\underline{X} = \{\underline{X}_1 ,\underline{X}_2 ,\ldots,\underline{X}_m\}$, $\underline{X}_i = \{a_1 ,a_2 ,a_3 ,a_4 ,a_5 ,a_6\}$, i=1,M
and M is a number of members.

Further with the definition of the object function Φ as the weight
or the volume of the structure, it leads to the quadratic form :

$$\Phi m = \Phi m(\underline{X},\underline{P}) = \frac{1}{2} * \underline{X}_m^T * A_1 * \underline{X}_m + \underline{X}_m^T * \underline{A}_2 + A_3 \tag{2}$$

where A_1 is a matrix, $\underline{A}_2$ is a vector, A_3 is a scalar, and they are
functions of the cross sections parameters. The function (2) is
called the global object function.

An admissible set of solutions G is composed by:

(i)   a statics nonlinear matrix equation :

$$K(\underline{X},\underline{B}_i) * \underline{B}_i = \underline{L}_i \qquad i=1,Lstb \tag{3}$$

where Lstb is a number of loading, $\underline{B}_i$ is a displacement vector
of joints and $\underline{L}_i$ is a external loading vector,

(ii)   a displacement constraints :

$$\underline{B}_i \leq \underline{B}_d \qquad i=1,lstb \tag{4}$$

230

(iii) a geometrical inequalities :

$$\underline{X}_L \leq \underline{X} \leq \underline{X}_U \qquad\qquad (5)$$

where $\underline{X}_L$, $\underline{X}$, $\underline{X}_U \in E$, and E is the finite
linear vector space,

(iv)   a stresses constraints gets from the Polish Steel Standard
written in form:

$$F(\underline{X},\imath) := (|\sigma(\underline{X},\imath)|/ R)-1 <= 0 \qquad\qquad (6)$$

Let $G_\imath$ denotes admissible set of the solutions for $\imath$ - state of
loading. Then G is equal :

$$G := \bigcap_{\imath=1}^{Lstb} G_\imath \qquad\qquad (7)$$

With the definition of the optimal criterium as the minimum, the
optimization model has the form [9]:

$$\Phi (\underline{X}^\bullet) = \underset{\underline{X} \in G}{MIN} \; \Phi(\underline{X},\underline{P}) \qquad\qquad (8)$$

Introducing additional file Ω^o defined as :

$$\Omega^o := \{ \; \underline{X} \in V^N \; : \; \Phi(\underline{X}) < \Phi(\underline{X}^\bullet) \; \} \qquad\qquad (9)$$

the necessary and sufficient condition for the existance the optimal
solution $\underline{V}^\bullet$ can be written as [8]:

$$\Phi (\underline{X}^\bullet) = \underset{\underline{X} \in G}{MIN} \; \Phi(\underline{X},\underline{P}) \; \leftrightarrow \; \Omega^o \cap G = \varnothing \qquad\qquad (10)$$

The "Back-Track" algorithm has been used [2], for looking the
optimal discrete solutions in optimization of frames expressed
by the equation (8).

2. The fundamental P-Δ approach.

The statics analysis of frames, expressed by the equation (3)
in optimization model, is considered for the small displacements,
linear-elastic condition, prismatic members, deterministic loading
with the influence of the geometrical nonlinearity [4].

These so called second order factors which has an influence on the internal forces distribution, are caused by displacements and deflections of a construction, and finally leads to increasing the value of internal forces with comparison to linear static analysis.

For example in two dimensional structural analysis, these increments are relate to shear forces and overturning moments but in three dimensional structures it concern to torsional moments also.

The overturning and torsional moments, acting at a given story of the structure have two components:

(i)   *primary moments received from the linear static analysis in which suitable differential or finite element equations are deviated of formula for undeformed state of structure,*

(ii) *second-order moments calculated for deformed state of the structure and are caused by the vertical forces acting over their respective incremental lever arms.*

This additional second order   components are called as the P-delta (P-Δ) effects [12].

For matrix displacement method, the P-Δ factors are introduced into linear elastic stiffness matrix for one member, before it transformation from the local to the global coordinate system. However for small deflection, i.e., when the influence of the local member curvature on the second order effects are negligible small, the P-delta effect can be introduced into matrix equilibrium equations after the assembly the global matrix K_s [10].

Let the matrix equilibrium equation has the form [6]:

$$K_s \cdot \underline{V} = \underline{F} \qquad\qquad (\ 11)$$

where $\underline{V} = \{\ \underline{V}_1, \ldots, \underline{V}_N\ \}$ is nodal displacement vector of construction, $\underline{V}_i$ is displacement vector for i-node (i.e. $= \{v_{ix},\ v_{iy}, o_i\}$ for plane structures), $\underline{F} = \{\ \underline{F}_1, \ldots, \underline{F}_N\}$ is lateral nodal loading vector and $\underline{F}_i$ is lateral loading acting on i-node (i.e. $= \{P_{ix}, P_{iy}, M_i\ \}$ for plane structures), N is a the total number of nodes.

Equations (11) is linear and present the standard linear method (LM). The P-Δ method modifies the right hand side of this equations by one factor. Let consider j-column and i-story of a rectangular structure, and assume that each story has l -columns. The P-Δ effects involved a second order moment M which have to be calculated at each frame story. This moment has the form [1, 4] :

$$M_i = (\ \sum_{j=1}^{l_j} N_{i,j}\ ) \cdot (v_{ix} - v_{(i-1)x}\ ), \quad i = 1, \ldots, m, \qquad (\ 12\ )$$

where $N_{i,j}$ is an axial force in j-column of i-story, v_i and v_{i-1} are horizontal displacements of the "i" and "i-1" stories and "m" is a number of stories.

Let define a lateral force couples F_{si} at each story as:

$$F_{si} = \alpha_i \cdot M_i / h_i \qquad\qquad (\ 13\ )$$

232

where h_i is a story height. Then lateral P-Δ vector has the form:

$$\underline{F}_s = \{ \, F'_{s1}, \ldots, F'_{sM} \, \}, \qquad F'_{si} = F_{s(i-1)} - F_{si} \qquad (14)$$

where "i" is a story number, M is a number of stories.
Making the transformation of the $\underline{F}_s$ vector, using the T matrix,
into the vector space for lateral loading $\underline{F}$, the influence of the
P-Δ effects in (11) can be express then in form :

$$K_s \cdot \underline{V} = \underline{F} + \underline{F}_d \qquad (15)$$

where $\underline{F}_d = T \cdot \underline{F}_s$.

3. Iterative solution for P-Δ method.

In the standard P-delta (SΔP) method, when the influence of
the axial force for deformation is neglected coefficient α_i is used
to take as equal 1 or 1.21 [3]. In modified (MΔP), it is calculated
from geometrical stiffness matrix [4], has an algebraic form and is
dependent only on the suitable rotations of nodes of stories for
linear model LM [12].It can be write as:

$$\alpha_i = f \, (\, \underline{V}_i \,) \qquad (16)$$

Hence and (15) the fundamental P-Δ matrix equations can be written:

$$K_s \cdot \underline{V} = \underline{F} + \underline{F}_d \, (\underline{V}) \qquad (17)$$

and its solution has an iterative form as follow :

(i)   first step $\qquad \underline{V}_0 = (\, K_s)^{-1} \cdot (\, \underline{F} \,) \qquad (18)$

(ii)   1-step $\qquad \underline{V}_1 = (\, K_s)^{-1} \cdot (\, \underline{F} + \underline{F}_{d1} \, (\underline{V}_{(1-1)})) \qquad (19)$

The iteration process stops when:

$$\| \, \underline{V}_1 - \underline{V}_{(1-1)} \, \| \, < \varepsilon \qquad (20)$$

where ε is a tolerance and $\|.\|$ is euclidean norm.
A convergence of SΔP method was considered in a papers [7],[14].
It can be found that for some structures, the process over five
iterations, is not convergent and structure could loose its global
stability.

4. Generalization of the P-Δ iterative procedure.

For small displacements under linear - elastic conditions, prismatic
members with neglecting the share shape deformation, the stress-
strain relations for construction can be written in a form [15]:

$$K_T(\underline{V}) \cdot \underline{V} = \underline{F}^n - \underline{F}^e(\underline{V}) \qquad\qquad (21)$$

where $K(\underline{V})$ is tangent stiffness matrix nonlinear dependent on a displacement vector $\underline{V}$, $\underline{F}^n$ is the loading vector acting on the nodes, $\underline{F}^e(\underline{V})$ is loading vector acting on the elements.

This equations can be solve by incremental iterative technique, which steps are as follow :

(i) first step
$$\underline{V}_0 = (K)^{-1} \cdot \underline{F}^* \qquad\qquad (22)$$

next are calculated
$$\underline{F}_0 := K_T(\underline{V}_0) \cdot \underline{V}_0 \;,\; \underline{F}^* := \underline{F}^n - \underline{F}^e \qquad\qquad (23)$$

$$\Delta\underline{F}_1^* := \underline{F}_0 - \underline{F}^* \qquad\qquad (24)$$

$$\Delta\underline{V}_1 = (K_T(\underline{V}_0))^{-1} \cdot \Delta\underline{F}_1 \qquad\qquad (25)$$

$$\underline{V}_1 = \underline{V}_0 + \Delta\underline{V}_1 \qquad\qquad (26)$$

(ii) i - step
$$\underline{F}_{(i-1)} := K_T(\underline{V}_{(i-1)}) \cdot \underline{V}_{(i-1)} \qquad\qquad (27)$$

$$\Delta\underline{F}_{(i-1)} := \underline{F}^* - \underline{F}_{(i-1)} \qquad\qquad (28)$$

$$\Delta\underline{V}_i = (K_T(\underline{V}_{(i-1)}))^{-1} \cdot \Delta\underline{F}_{(i-1)} \qquad\qquad (29)$$

$$\underline{V}_i = \underline{V}_{(i-1)} + \Delta\underline{V}_i \qquad\qquad (30)$$

All process is finished when :

$$\| \Delta\underline{F}_{(i-1)} \| < \varepsilon \qquad\qquad (31)$$

for small arbitrary chosen $0 < \varepsilon \ll 1$.

This iterative increment procedure direct leads to the secant process (NMII), in which i-iteration is described as :

$$\underline{V}_i = K_T(\underline{V}_{(i-1)})^{-1} \cdot \underline{F}^* , \qquad\qquad (32)$$

what has been presented on figure 2.

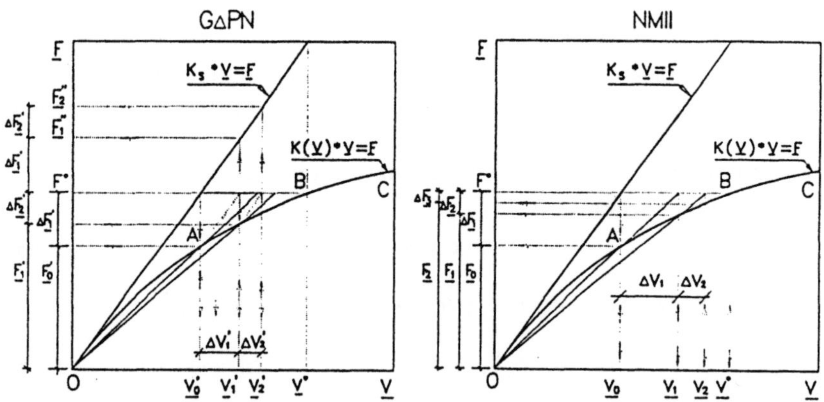

Fig. 2. General P (G PN) and standard secant stiffness
       (NMII) approaches for the II order theory.

Now let assume that on each iteration in (32), the inverse matrix
will be calculated from elastic matrix K_s . Then from (32) :

$$K_s \cdot \underline{V}_i = \underline{F}^\bullet + (K_s - K_T(\underline{V}_{(i-1)})) \cdot \underline{V}_{(i-1)} \qquad (33)$$

This equation is called the generalization of the P-∆ method.
It can be note that it has similar form then equation (17).
After introducing incremental loading defined as :

$$\underline{F}_i^{''} = \underline{F}_{(i-1)}^{''} + \Delta \underline{F}_i \qquad (34)$$

where
$$\Delta \underline{F}_i = \underline{F}^\bullet - \underline{F}_{(i-1)} \qquad (35)$$

equation (34) can be written as :

$$K_s \cdot \underline{V}_i = \underline{F}^\bullet + \sum_{j=0}^{i} \Delta \underline{F}_{(j-1)} \qquad (36)$$

which give the sense of the P-∆ effects in general nonlinear
expression (G∆PN).

When the global stiffness matrix K_T is linearized by elastic matrix
K_s and geometric matrix K_G i.e.:

$$K_T \simeq K_s + K_G \qquad (37)$$

then substituting this expression into (35), it can be obtained
a linearized general P-∆ method (G∆PL), which for i iteration has
the form:

$$K_s \cdot \underline{V}_{1} = \underline{F}^{\bullet} + K_G (\underline{V}_{(1-1)})) \cdot \underline{V}_{(1-1)} \qquad\qquad (38)$$

It can be noted that both GΔPN and GΔPL methods are convergent.

5. Numerical example

As an example of an effectivity of the discrete optimiztion model with the comparison of all, P-Δ approaches for the LM, MNII and MNIIL, it has been choossen simple plane frame shown on a figure 3.

In this example ,the V_{1x} displacement, for x direction in joint number 1 and the M_{12} internal moment in joint number 12, are taken under consideration.

There are two groups of materials and cross sections for the frame. One is composed by the exterior columns, the second by the rest of members. Ich group has double symmetric T section with all variable dimensions so the design vector $\underline{X}$ is six dimensional.

The initial values for $\underline{X}$ vector are {5,1,1,5,5,1,1,5} and iterations step is equal 0.5. Lower and upper geometrical bounds has been taken as follow : $\underline{X}_L$ = { 3,0.5,0.5,4,3,0.5,0.5,4},
$\underline{X}_U$ = { 6,1,1,6,6,1,1,4}

and the maximum number of iteration was assummed as 6.

The object function was defined as the volume of the frame.
It has been introduced a relative error defined as [5]:

$$\varepsilon_{M,X} = \frac{\| w - w' \|}{w} \cdot 100 \ \% \qquad\qquad (39)$$

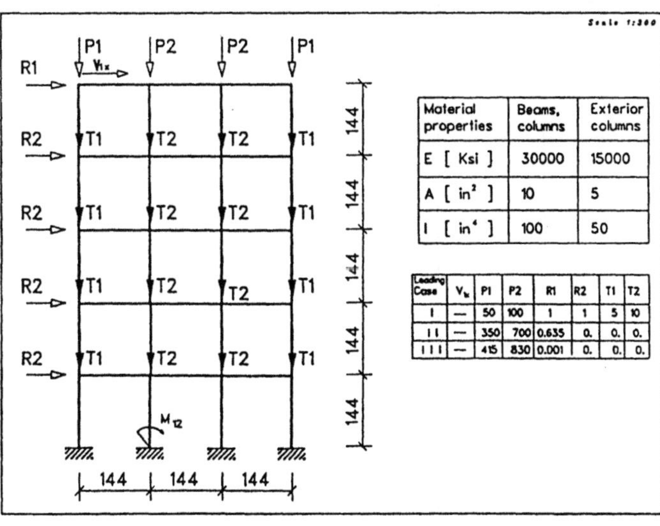

Material properties	Beams, columns	Exterior columns
E [Ksi]	30000	15000
A [in²]	10	5
I [in⁴]	100	50

Loading Case	V_x	P1	P2	R1	R2	T1	T2
I	—	50	100	1	1	5	10
II	—	350	700	0.635	0.	0.	0.
III	—	415	830	0.001	0.	0.	0.

Fig. 3. Topology, geometry, material properties and loading of a three-bay, five-storey frame.

236

where w is a solution received by the MNII method and w' is a calculated value by the rest methods.

This relative error gives the information about effectivity of the one method (in an accuracy sense of the one value) due to accurate solution.

In this example the time of the computer operations has been measured also. For this purpose the relative coefficient has been defined :

$$c_T = t_M / t_{LM} \qquad\qquad (40)$$

where t_M , t_{LM} are times of calculations for chosen and ML method respectively. This definition gives the proper look for an effectivity of one method because the direct calculation time is in great manner perturbed by the chosen compiler, linker and the type of a computer.

All calculation has been made under FOPT_PC computer program written in MS FORTRAN v.5.0, MS C v.6.0 and MASM v5.0, on the base of PC 486/25MHz/EVGA unit.

With the use of the "Back-Track" technique of the solution the FOPT_PC program has founded the optimal solution almost at the boundary, for all cases in point $X^* =\{ 3,1,1,4,3,0.5,0.5,4\}$. The object function has been changed its value from 75600 in^3 to 36000 in^3 which occured the optimal.

In a table 1 are presented results for loading case I, and loading case III is in the critical area. This critical loading is R1=0 [kips] and P= 830[kips] for NMII .

Table 1

Calculated displacement, moment relative errors		LM	$S\Delta P_{\mathcal{L}=1.21}$	$M\Delta P_{\mathcal{L}=1.60}$	GΔPL	NMIIL	GΔPN	NMII
I	V_{tx} [in]	0.7634	0.8864	0.8869	0.8878	0.8909	0.8877	0.8910
	ε_x [%]	14.32	0.51	0.46	0.36	0.01	0.28	0.0
	M_{12} [kip·in]	−136.9456	−139.4758	−148.5531	−153.5158	−153.9283	−153.4749	−153.9069
	ε_M [%]	11.02	9.38	3.48	0.25	0.01	0.28	0.0
	ε_T [%]	1.	2.41	2.86	3.01	3.44	3.6	4.16
III	V_{tx} [in]	0.00026	0.00030	0.00030	0.00033	0.06307	0.00033	0.23069
	ε_x [%]	99.89	80.68	80.87	83.21	1.18	83.24	0.0
	M_{12} [kip·in]	−0.0237	−0.02419	−0.02577	−0.02788	−2.26497	−0.02789	−8.00549
	ε_M [%]	99.7	99.87	99.87	99.86	72.66	99.86	0.0
	ε_T [%]	1.0	2.72	3.23	3.4	3.9	4.11	4.77

It has been assumed that ε, which ends the iterations, is equal to the value 1E-5.

For each step of the optimization it has been the four iterations for the solutions of nonlinear statics analysis equation.

Finally received results shows that:

(i) In non critical area standard and modified P-Δ methods gives relative error in the range of 3.5 - 9.5 %, but general P-Δ methods are much more accurate with relative errors in the range of 0.25-0.28 % .

(ii) For of the relative time coefficient, the differences between GΔPL and MΔP are small (relative range $\approx$ 2.4%), but between GΔPN and MΔP are considerable (relative range $\approx$ 20.3%).

(iii) In critical area of loading, within the four iterations, only the MNII method has given the results near the analytical solution. The MNIIL with relation to the MNII has given 1.18 % error but all P-Δ have given this error huge, in a range of 80.7 -83.2

(iv) Increasing iterations untill 50, the MNII has oscillated around value received in fourth iteration (table 1), MNIIL, GΔPN and GΔP has been convergent to solutions of MNII method, but SΔP and MΔP methods have not been convergent at all.

6. Conclusions

The general P-Δ approaches, described in this paper, are more effective then the standard and modified P-Δ methods.

It is not the consequence of the direct comparison of these methods on the base of a rectangular frames only but the general form of the matrix equation [33,36].

This form :

(i) *doesn't require any special topology and geometry of the structure, so any structure can be consider,*

(ii) *is convergent if the MNII and MNIIL methods are convergent*

(iii) *has general form what caused that all second order factors are considered, not only those mentioned in SΔP and MΔP methods.*

(iv) *is highly recomended for discrete optimization models of frames with opposite to MNIIN as fast, enough accurate and convergent even in critical area.*

238

References

1. Berkowski P., *Approximate second order analysis of unbraced multistory steel frames*, Proc. of the International Conference on "Metal Structures", Gdańsk, Poland, 1989.

2. Berkowski P., Boroń J., *Synthesis of Some Structures with the "BackTrack" Discrete Programming Method"*, (in polish),Inz. i Bud., 43/86, s.204-208.

3. Berkowski P., *Discrete Optimization of Highrise Steel Frames*, International Conference on Highrise Buildings, Nanjing'89, China.

4. Birnstiel C., Iffland J.S., *Factors Influencing Frame Stability*, J. Struct. Div., Proc.ASCE, 106, ST2, Feb. 1980, 491-504,

5. Fellippa, C.A., *Error Analysis Of Penalty Function Techniques For Constraint Definition In Linear Algebraic Systems*, Inter.J.Num.Meth.Eng.,Vol 11, 1977,709-728,

6. Gaiotti R., Smith B. S., *P-delta analysis of building structures*,J. Strict. Eng.,Proc. ASCE,4, ST115, 1989, 755-770,

7. Ilków W., *Accuracy of the P-Δ method*, T5, Warszawa 1983, IPPT PAN,189-202,

8. Janczura T. Andrzej, Sieczkowski J., *Optimization of the Frame Construction Using Computers*,(in polish), Inzynieria i Budownictwo, Nr 4-5, 1986, pp. 139-142,

9. Janczura, A.T., *Optimization of the Structures With the Use Of The Network Method*, Phd.thesis,(in polish), Building Engineering Institute, Technical University of Wroclaw, May 1982,

10. Janczura Andrzej T., *The Network Approach in the Finite Elements Method*,"Computer and Mechanics, Poland, T9, Warsaw, 1989, 241-256,

11. Lai Shu-Ming A., MacGregor J.G., *Geometric Nonlinearities in Unbraced Multistory Frames*, J. of Struct. Eng., Proc. ASCE,109, ST11, Nov. 1983, 2528-2545,

12. Neuss C.F., Maison B., *Analysis for P-Δ Effects in Seismic Response of Buildings*, Comp. & Struct., vol. 19, 3, 1984,

13. Rutenberg A., *A Direct P-Delta Analysis Using Standard Plane Frame Computer Programs*, Comp. & Struct., vol. 14, 1-2, 1981,

14. Wood B.R., Beaulieu D., Adams P.F.,*Column design by P-Delta method*, J. Struct. Div.,Proc. ASCE, 102, ST2,1976, 411-427,

15. Zienkiewicz O.C., Cheung Y.K., *The Finite Elements Method in Structural and Continum Mechanics*, Mac Graw-Hill Book Comp. ,New-York,1967.

A More General Optimization Problem for Uniquely Decodable Codes

Ahmed A. Belal

Department of Computer Science, Faculty of Engineering, University of Alexandria, EGYPT

Abstract: The problem of finding the integers $r(i)$, $i = 1, 2, \ldots, N$ that minimize the cost function $J = \sum_{i=1}^{N} A(i)\, f[r(i)]$ under the constraint $\sum_{i=1}^{N} b^{-r(i)} \le b^{t}$ where t is a non-negative integer and b is an integer ≥ 2, is shown to have a very simple solution when the function $f[r]$ is an exponential function of r and $A(1), A(2), \ldots, A(N)$ is any set of positive real numbers.

It is shown that when $f[r] = a^{r}$ where a is any real positive number, a simple procedure based on the theory of uniquely decodable compact codes can be used to determine the integers $r(1), r(2), \ldots$ when $t = 0$. A minor modification on the already obtained values for the $r's$, will allow for other values of t.

1- INTRODUCTION

It is well known from the theory of uniquely decodable codes [1], [2], [3] that the lengths $r(1), r(2), \ldots, r(N)$ of the code words must satisfy the condition $\sum_{i=1}^{N} b^{-r(i)} \leq 1$ where b

is the size of the code alphabet.

In the case of binary codes, $b = 2$, the condition becomes $\sum_{i=1}^{N} 2^{-r(i)} \leq 1$.

$$(1)$$

In this case, given any set of positive real numbers A (1), A (2), ..., A (N) the integers r (1), r (2), ..., r (N) satisfying equation (1) and minimizing the cost function $J = \sum_{i=1}^{N} A(i) \, r(i)$ will be the set of word lengths of a compact code obtained from a complete code tree of order two, and satisfying equation (1) with equality [1], [4].

For N = 2 and N = 3 there is only one complete binary code tree giving the solution sets {1,1} and {1, 2, 2} respectively. For N > 3 more than one solution set exists. Fig. 1 shows the three possible trees for N = 5, giving the solution sets {2, 2, 2, 3, 3}, {1, 3, 3, 3, 3}, {1, 2, 3, 4, 4}.

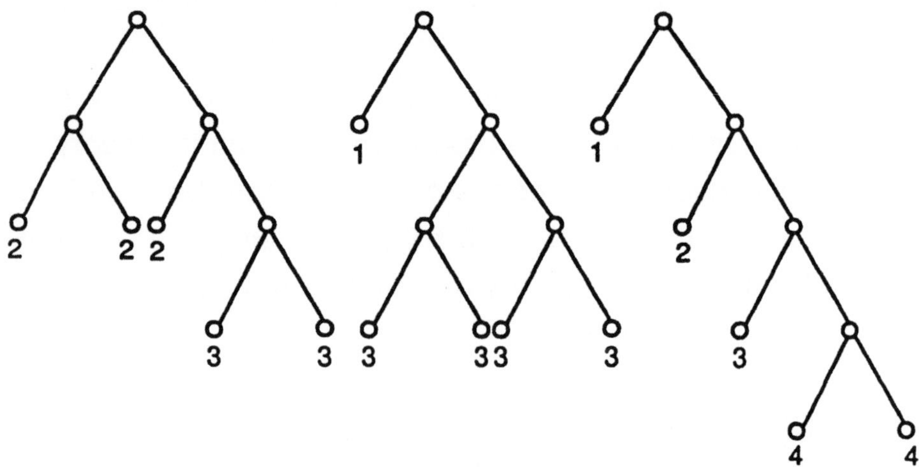

Fig. 1

It is easy to see that the solution sets obtained for binary compact codes will also minimize the cost function

$$J = \sum_{i=1}^{N} A(i) f[r(i)]$$

subject to the constraint of equation (1) for any function satisfying the condition $f[r(i)] \geq f[r(j)]$ for $r(i) \geq r(j)$ by simply assigning the smaller $r(i)$ to the larger $A(i)$.

In the next section, it will be shown that a Huffman like procedure for compact codes [2], [3] can be used to uniquely determine the solution set $r(1), r(2), ..., r(N)$ when f is the exponential function $f[r(i)] = a^{r(i)}$ where a is any positive integer greater than 1.

Extensions to non-binary cases when $b > 2$ will also be given.

2.1 The Binary Case b=2

Without any loss of generality, assume that the real positive coefficients $A(i)$ are such that $A(1) \geq A(2) \geq A(3) \geq .... \geq A(N-1) \geq A(N)$.

Let $r(1), r(2),.....,r(N)$ be the set of integers that minimizes the cost function $J = \sum_{i=1}^{N} A(i) f[r(i)]$, under the

constraint $\sum_{i=1}^{N} 2^{-r(i)} \leq 1$, when the function f satisfies the

condition

$$f[r(i)] \geq f[r(j)] \text{ when } r(i) \geq r(j) \tag{2}$$

From the properties of the complete binary code tree, it is clear that $r(N-1) = r(N)$.

We now form a reduction of the problem by replacing the two coefficients $A(N-1)$, $A(N)$ with the coefficient

242

$$A(0) = a[A(N-1) + A(N)] \qquad (3)$$

where a is any positive real number.

We would like now to solve the reduced problem and find the (N-1) integers r' (1), r' (2),.....,r' (N-2), r (0) that will minimize the cost function.

$$J' = A(0)\, f[r(0)] + \sum_{i=1}^{N-2} A(i)\, f[r'(i)] \qquad (4)$$

under the constraint $2^{-r(0)} + \sum_{i=1}^{N-2} 2^{-r'(i)} \leq 1$ (5)

then use the same binary code tree obtained, to get the solution to the original problem by just adding two branches to the node representing r (0). Thus the solution set to the original problem will be

r (i) = r' (i) for i = 1, 2,.....,N-2

$$r(N-1) = r(N) = r(0) + 1 \qquad (6)$$

In order to be able to do this, the two cost functions J of the original problem and J' of the reduced problem must differ by at most a constant value.

From equations (3), (4), (6) we obtain

$$J - J' = \{A(N-1) + A(N)\}\, \{f[r(0) + 1] - a\, f[r(0)]\} \qquad (7)$$

Setting his difference equal to a constant irrespective of the value of r (0) gives the difference equation

$$f[r+1] - a\, f[r] = k \qquad (8)$$

where k is any constant value.

Equation (8) has the two solutions

$$f[r] = kr \qquad \text{when } a = 1 \qquad\qquad (9a)$$

$$f[r] = a^r \qquad a \neq 1, k = 0 \qquad\qquad (9b)$$

Solution (9a) produces the Huffman encoding method where the cost function is the average code length.

The other solution (9b) shows that the simple procedure outlined can be used to optimize the cost function

$$\sum_{i=1}^{N} A(i)\, a^{r(i)} \qquad\qquad (10)$$

Since this last solution was obtained with k=0, both cost functions J,J' will have identical values in this case.

Solution to the problem is found by making successive reductions, until we reach the trivial case of three variables having the one solution set {1, 2, 2}.

The following example shows the procedure for minimizing the cost function

$$J = 330\,(4)^{r(1)} + 70(4)^{r(2)} + 15\,(4)^{r(3)} + 10\,(4)^{r(4)} + (4)^{r(5)} + (4)^{r(6)}$$

$$\text{subject to the constraint} \quad \sum_{i=1}^{6} 2^{-r(i)} \leq 1$$

Here A(1)=330, A(2)=70, A(3)=15, A(4)=10, A(5)=A(6)=1 and a=4.

First, we arrange the coefficients in descending order, then the first reduction is obtained by replacing the two smallest coefficients A (5), A (6) with one coefficient of value A (0) = a [A (5) + A (6)] = 8.

The coefficients are again arranged in descending order as in the second column. The process is repeated until we reach the final stage with the three coefficients 340, 330, 72 for which the solution set is {1, 2, 2}.

Moving backwards we obtain the solution set $\{2, 2, 2, 2\}$ for the stage with coefficients 330, 72, 70, 15 then the set $\{2, 2, 2, 3, 3\}$ for the stage with coefficients 330, 70, 15, 10,8 and finally the solution $\{2, 2, 2, 3, 4, 4\}$ for the original problem.

We note that the cost function $\sum_i A(i)\, 4^{r(i)}$ has the same value of 7792 in all the stages.

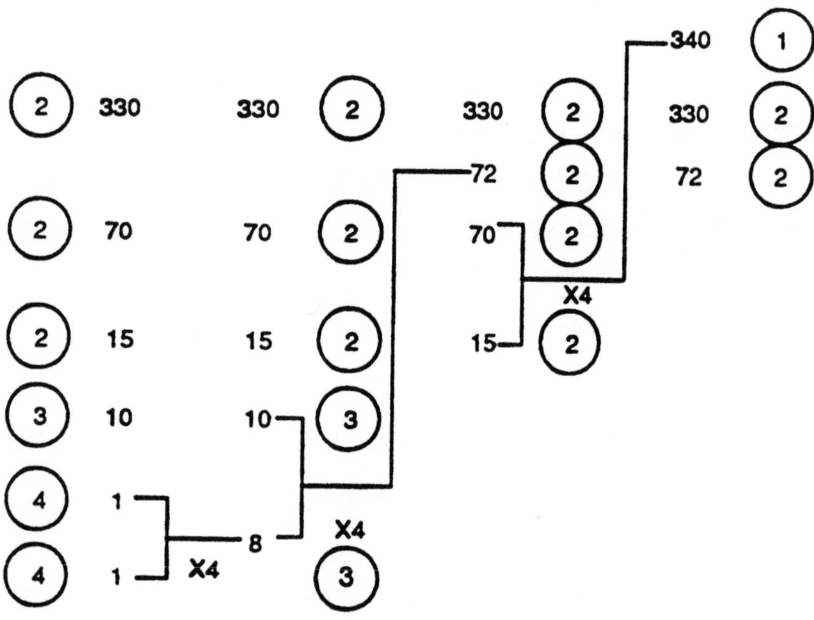

Solution sets for several different coefficients are given below for the case $f[r] = 4^r$. The first entry corresponds to the previous example.

Coefficients A (1), A (2).......A (N)	Solution set
330, 70, 15, 10, 1,1	{2, 2, 2, 3, 4, 4}
349, 75, 15, 10, 1, 1	{1, 2, 3, 4, 5, 5}
42, 8, 6, 6, 4, 2, 1	{2, 3, 3, 3, 3, 3, 3}
1, 1, 1, 1, 1, 1, 1, 1	{3, 3, 3, 3, 3, 3, 3, 3}
128, 64, 32, 16, 8, 4, 2, 1	{2, 2, 3, 3, 4, 4, 4, 4}
65, 15, 8, 8, 6, 4, 1, 1, 1	{2, 3, 3, 3, 3, 4, 4, 4, 4}
3841, 200, 100, 76, 60, 40, 8, 1,1	{1, 3, 3, 4, 4, 4, 5, 6, 6}
3839, 200, 100, 76, 60, 40, 8, 1,1	{2, 3, 3, 3, 3, 3, 4, 5, 5}

The given procedure will minimize the cost function J, provided that the function f satisfies the condition given in equation (2) namely, that f [r] increases when r increases. Thus for the exponential function a^r, the value of a must be greater than 1.

For values of a less than 1, the previous procedure will maximize the cost function J under the constraint $\sum_{i=1}^{N} 2^{-r(i)} = 1$.

The following example gives the solution set {1, 2, 3, 4, 4} to maximize the cost function

$$J = 9 (3/4)^{r(1)} + 8 (3/4)^{r(2)} + 5 (3/4)^{r(3)} + 4 (3/4)^{r(4)} + 4 (3/4)^{r(5)}$$

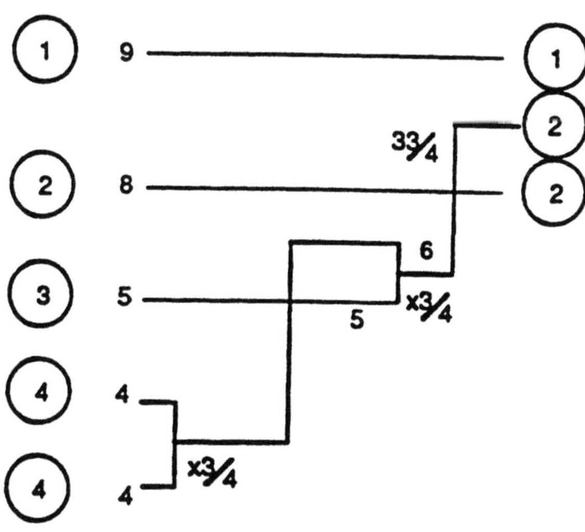

The case a ≤ 1/2 is of special interest, since in this case the new coefficient A (0) = a [A (N-1) + A (N)] that replaces the two smallest coefficients A (N-1), A (N) will always be smaller or equal to the coefficient A (N-2).

With this condition satisfied in every reduction, the only possible solution set when a ≤ 1/2 is the set

$$r (k) = k \qquad\qquad , k = 1, 2, ..., N\text{-}2$$
$$r (N\text{-}1) = N\text{-}1 \qquad\qquad\qquad (11)$$
$$r (N) = N\text{-}1$$

Since the solution set always forms a complete code tree satisfying condition (1) with equality, the given procedure can still be used to minimize the cost function J under the constraint $\sum_{i=1}^{N} 2^{-r(i)} = 1$, when the value of the exponent a is less than 1, by simply assigning the bigger r (i) to the bigger coefficient A (i). This is achieved by arranging the coefficients A (1), A (2), ..., A (N) in ascending order then forming a reduction by replacing the largest two coefficients A (1), A (2) with the coefficient A (0) = a [A (1) + A (2)].

The following example demonstrates the procedure to minimize the cost function J under the condition $\sum_{i=1}^{6} 2^{-r(i)} = 1$, with the coefficients 45, 36, 30, 24, 21, 16 and the function $f [r] = 3^{-r}$

The solution {2, 2, 2, 3, 4, 4} gives the minimum cost function

$$J(\min)=16\,(3)^{-2}+21\,(3)^{-2}+24\,(3)^{-2}+30\,(3)^{-3}+36\,(3)^{-4}+45\,(3)^{-4}=80/9$$

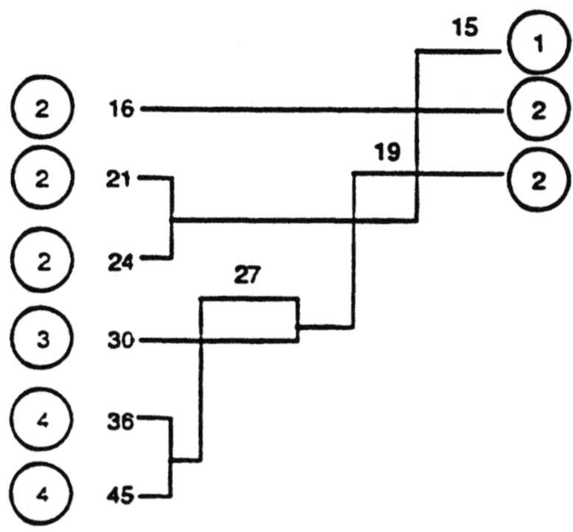

2.2 Extension to the Non-Binary Case b > 2

When b is greater than 2, the same results of the previous section still apply provided that the number of coefficients N can form a complete code tree of order b [1], [5].

A complete code tree of order b, is one in which every node has b branches coming out of it. It is clear that the number of leaves N in such tree must satisfy the condition

$$N = b + (b - 1)\,k \quad \text{where } k = 0, 1, 2,.... \tag{12}$$

By reducing the number of variables by (b-1) in every step, the process will end after k steps leaving a problem with b variables having the solution set $r\,(1) = r\,(2) = r\,(3)$ =....... = $r\,(b) = 1$ and satisfying the condition $\sum\limits_{i=1}^{N} b^{-r(i)} = 1.$

When the number of given coefficients N, does not satisfy equation (12), additional coefficients having value equal to zero are added so that (12) is satisfied.

Finally, if the constraint of equation (1) is of the form

$$\sum_{i=1}^{N} b^{-r(i)} \leq b^{t}$$

where t is an integer, then the optimization procedure will give the solution set for the integers $r(i) + t$.

CONCLUSION

It was shown that the theory of uniquely decodable codes can be used to optimize a more general cost function than the average length of the code.

The problem of optimizing the cost function $J = \sum_{i=1}^{N} A(i) f[r(i)]$, under the constraint $\sum_{i=1}^{N} b^{-r(i)} \leq b^{t}$, was solved when $f[r]$ is the exponential function a^{r}.

It was proved in this case, that the value of the cost function remains unchanged under a reduction of the problem from N variables to (N-b+1) variables.

REFERENCES

[1] Gallager, R.G., "Information Theory and Reliable Communication", Wiley, 1968.

[2] Abramson, N., "Information Theory and Coding", McGraw-Hill, 1963.

[3] Ash, R., "Information Theory", Interscience, 1965.

[4] Knuth, D.E., "Sorting and Searching", Addison-Wesley, 1973.

[5] Aho, Hopcroft, Ullman, "The Design and Analysis of Computer Algorithms", Addison-Wesley, 1974.